G. Fandel H. Dyckhoff J. Reese (Eds.)

Essays on

Production Theory and Planning

With 48 Figures
and 46 Tables

Springer-Verlag
Berlin Heidelberg New York
London Paris Tokyo

Prof. Dr. Günter Fandel
FernUniversität Hagen, Fachbereich Wirtschaftswissenschaft
Lehrstuhl für Produktion und Investition
Feithstr. 140, D-5800 Hagen 1

Prof. Dr. Harald Dyckhoff
RWTH Aachen, Lehrstuhl für Industriebetriebslehre
Templergraben 64, D-5100 Aachen

Prof. Dr. Joachim Reese
Rheinische Friedrich-Wilhelms-Universität Bonn
Betriebswirtschaftliche Abteilung
Adenauerallee 24–42, D-5300 Bonn

ISBN-13:978-3-642-73750-3 e-ISBN-13:978-3-642-73748-0
DOI: 10.1007/978-3-642-73748-0

Library of Congress Cataloging-in-Publication Data.
Essays on production theory and planning / G. Fandel, H. Dyckhoff, J. Reese, editors. p. cm.
Bibliography: p.
ISBN-13:978-3-642-73750-3
1. Production management. 2. Materials management. I. Fandel, G. (Günter), 1943– .
II. Dyckhoff, H. (Harald) III. Reese, J. (Joachim) IV. Title: Production theory and planning.
TS155.E77 1988 658.5–dc19 88-19899 CIP

Softcover reprint of the hardcover 1st edition 1988

2142/3140-543210 Printed on acid-free paper

To

Gabriele, Birgit, and Jutta

Preface

The thirteen essays of this book deal with various aspects of production management which have shown a growing importance in research, teaching and practice within the last years and are partly neglected in the literature. The contents range from theoretical issues to case studies. Significant relations among different essays (chapters) give rise to group them into four major parts dealing with separate themes of production management:

- fundamental organizational aspects of the production process,
- general problems and modern concepts of materials management,
- joint production with surplus, waste and hazardous byproducts,
- production aspects of cutting stock and trim loss problems.

Moreover, there is a strong material connection between the contributions to this volume because they all are based on a common conceptual framework emphasizing the production theoretic foundation.

The collection of papers in this book is also designed to provide a fairly comprehensive picture of scientific work in the field of production management done during the last years at the Institute of Production Economics of the Fernuniversität in Hagen, partly in cooperation with other scientists. Since all but two articles have not been published in English before, the reader gets quite a new view on the results of this recent research.

Part I of the volume considers some fundamental organizational aspects of the production process. The need for good manufacturing organization is demonstrated in *Chapter 1*. As the reader will learn, the choice of the organizational structure of the manufacturing process has deep consequences for the subsequent planning procedures as a whole. From the discussed basic structures, i.e. the machine oriented line organization and the product-oriented process organization, it is concluded, how coordination efforts alter. As is further shown, planning objectives and planning procedures must be adopted to these different organizational decisions. In turn, if concrete objectives were required with respect to the flowtimes of jobs or the available machine capacities, this means that a suitable organization structure is to be chosen. *Chapter 2* extends these organizational reflections onto the logistic system within the firm. By means of a case study for the vehicle utilization the interdependencies between organizing and planning the optimum system is exemplified. Of particular interest in this context is that uncertainties arise, which are due to the effective order volume and therefore require modified optimization methods. In spite of long-term organizational decisions, which remain unchanged during the whole planning period, a short-term planning procedure with a high degree of flexibility is installed so that remarkable gains occur as compared with actual results. Hereafter, in *Chapter 3*, another object of organizational interest is concerned, namely the purchasing sector. Materials management, and especially the purchasing of goods, is often discussed as to its optimal degree of decentralization. Within this framework, an answer is tried with regard to the inherent order policy that is applied. For the one-item case and independent demand rates at the various manufacturing plants, where the item serves as a production input, it is developed that some order policies favour centralized purchasing activities, whereas other policies support the establishment of a decentralized purchasing management.

Part II is dedicated to more general problems and modern concepts of materials management. *Chapter 4* is mainly concerned with a critical assessment of the popular "materials requirement planning" (MRP) and the Japanese "Kanban"-system. Again, different organizational issues of both concepts restrict their efficient applicability to a special class of material planning problems. MRP proposes a centralized, multi-stage mechanism, which includes an analytic bill explosion as well as an optimum lot sizing procedure. On the other hand, Kanban describes a simple, but nevertheless effective possibility of decentralizing that process by a retrograde automatism. Then, regarding the distinct coordination and information necessities, some most striking results are derived. Another modern concept of material handling, which sweeps through whole industries, is that of just-in-time production. It is examined in *Chapter 5* from the view of a supplier, when he is confronted with special requirements by the customers, i.e. their order quantities and delivery dates. As is evident, the burden of carrying inventory costs is shifted onto the supplier and thereby influences his optimal lot sizes in the production schedule. The policy, which minimizes the sum of inventory and set-up costs, is discussed in a theoretical analysis, using the economic order quantity (EOQ)–rule. Subsequently, the results are visualized by applying them in a case study. It is shown to what degree a customer must change the time-splitting of his orders, before a negative consequence is recognized for the supplier's production schedule. Finally, a contribution in this part deals with a single, but undoubtedly most important production input which cannot be added simply to the discussion of other materials and therefore requires singular calculation methods. Forms of this input can be summarized as "energy". *Chapter 6* deals with a special aspect of energy planning, namely the smoothing of gas and current purchase in industrial companies. Various methods on the basis of demand interruptions are compared. For the less familiar conception of making up machine schedules a decision model is formulated and widely interpreted. Its applicability is demonstrated for two practical industrial cases.

Some of the considerations on economic energy processing of Chapter 6 are exemplified by the case study for a small-scale power station at the beginning of Part III. *Chapter 7* deals with the handling of constant power-heat-relations resulting from a firm's energy generation for own purposes. It is shown that the problem can be expressed by a linear programming formulation. The fundamental description of the observed case is as follows: Steam is generated alternatively in three boilers and, subsequently, transformed into power as well as into heat by turbines. The constant relations of the respective output quantities depend obviously on the type of turbine involved in the energy generation process. Surplus production of heat is not feasible, as the quantities, which exceed the periodical demand, cannot be emitted due to air pollution acts. This is why the planning staff must be engaged in solving this problem of joint production by optimum machine employment. The influence of the choice done on the resulting energy costs is demonstrated. Furthermore, additional market transactions with regard to the generated quantity of electric power often are inevitable. This is due to the joint production nature of the energy generation process with power and heat as utilizable co-products. In general, however, by-products of energy generation can also be non-utilizable to the producer or even hazardous, e.g. radioactive substances such as plutonium. In this connection, technological risk as well as public and environmental health considerations have a decisive influence on the performance of industrial processes. *Chapter 8* refers to a particular case study from chemical industry. For a plant presenting the so-called SEVESO-case (chlorination of phenol with by-product dioxin) a mathematical model is developed. The integrated production planning under the prospect of hazardous by-products leads to a multiple criteria decision making problem solved by a nonlinear programming method within a concept for decision support. In contrast to the two prior case studies, the last chapter of Part III considers the problem of undesired by-products in joint production from a more general, theoretical point of view. *Chapter 9* is concerned with the question to what extent the optimal design of a production program is affected by variable disposal costs and maximum disposal quantities for those products that can neither be used within the production process nor marketed. Both are integrated as parameters into a linear programming model which may be regarded as a synthesis of most planning approaches in joint production which can be found in the (German-language) literature so far, and which contains the latter as special cases for the most part. By means of a numerical example it is demonstrated what changes in profits will possibly have to be put up with, depending on which special case is chosen for the formulation of the problem, and in what way the surplus quantities of individual joint products will then behave. The results obtained are directly applicable to a practical planning situation, since they show that deliberately allowing for surplus quantities under profit aspects is still advantageous, even if comparatively high disposal costs have to be taken into account – which at first sight could suggest a general avoidance of surplus quantities.

Part IV concentrates on production aspects in connection with the so-called "cutting stock" or "trim loss" problems. However, because of a strong relationship to packing and loading as well as other geometric assignment problems, the contents of this part are of importance for other areas of production management, too. In particular, elastic joint production is addressed thus extending the different investigations on (rigid) joint and waste production of Part III. *Chapter 10* develops the fundamentals of a production theory of cutting (stock) and related processes based on an activity analytic approach. The systematic analysis of a few basic assumptions allows production models to be deduced which normally form the starting point in the application oriented literature. Thus the underlying premises of these models become apparent. Therefore prerequisites are created which make the relations between different model approaches easier to recognize, as well as the requirements for the utilization of these approaches in practice. *Chapter 11* illustrates this for a concrete example of a trim loss and inventory planning problem in a small textile firm: A stock of rolls of a synthetic cloth has to be cut such that the sum of net returns and the value of the residual stock at the end of the planning period is maximized. Two linear optimization models are compared. Since a limited number of "good" cutting patterns is heuristically chosen for the first model it yields suboptimal solutions only. On the contrary, the second model takes all relevant cutting patterns into account and provides an optimal solution. In general, however, the efficiency of the various solution approaches to cutting stock problems is very problem dependent. Most of them are heuristics strongly based on the particular problems to which they are applied. Therefore the study of papers on problems with a comparable structure may prove helpful. In *Chapter 12*, a detailed catalogue of criteria for the characterization of real world trim loss (cutting stock) problems is developed in a differentiated and systematic manner. The main criteria are: dimension, planning situation, goals, restrictions, and solution approach. The proposed classification system is based on 38 case studies found in the literature. The classification of these case studies is represented by means of a "classification schedule". *Chapter 13*, the last chapter of the book, extends and modifies this classification scheme for a survey of standard software for the support and optimization of industrial cutting processes found by an empirical examination of the German software market. 28 program packages offered by a total of 15 firms are thus compared.

We wish to thank our colleagues who contributed to the book. We would also like to thank the Fernuniversität in Hagen for its generous support.

To Gabriele Dahl, Manuela Kutschinski, Lioba Lanwer, and Frank Reh we are indebted for their patient typing and retyping of the manuscripts. The drawings were done by Thomas Giersiepen and Ralf Schmiedgen. For all this help, so graciously provided, the editors express their thanks.

Hagen, April 1988

GÜNTER FANDEL
HARALD DYCKHOFF
JOACHIM REESE

Table of Contents

Part I

Organizational Aspects of Production

Chapter 1

Consequences of the Organizational Structure for the Production Planning System

JOACHIM REESE

I. Introduction

Firms which produce different goods in multistage production processes are faced with the important task of organizing these processes so that they are in line with the firm's objectives. Where the size of a firm and its channels of information prohibit the use of centralized decision making in short-run problems, a partial response to the organization problem will often be made by decentralizing responsibility for the job and machine schedules, respectively. Other important reasons for decentralizing responsibility are shown, for example, by MINTZBERG (1979, p. 182). They all stipulate that in many cases essential organizational tasks lie in deciding which persons should be made responsible for which parts of production planning.

In the past decades the theory of organization has developed many suggestions to solve these important questions (see e.g. HAX and MAJLUF 1981), suggestions which, in their basic forms, are valid for all areas of the firm, but which also determine a precise structure specifically for the area of production. The classical functional and the newer divisional types of structure are regarded as opposite poles between which all the other mixed structures are to be found. Whereas the functional type distributes responsibility on an input basis, e.g. with regard to the task performances, the divisional type is keyed to the purpose of production and is therefore output-orientated. The global advantages and disadvantages of these structures, and of all the others, can sometimes be compared quite easily and thus give grounds for a suitable organizational decision. But often more concrete statements are required as to how much different organizational structures can assist the planning system in conveying the objectives of the central authority and in utilizing the management's information as best and as quickly as possible, for example. A thorough examination of different organizational structures is called for, especially where those planning problems are concerned which often recur in similar form and which can usually be solved by routine procedures.

In the following two aspects will be dealt with which must not be ignored when choosing an organizational design for the production process: the handling of prefixed objectives within the organizational units and the effects of decentralization on the production planning system. Firstly it will be shown which consequences are to be expected from an organizational structure regarding the fulfilment of given objectives. Secondly the applicability of two decentralized heuristic planning procedures will be discussed inasmuch their inherent coordination efforts and information activities depend on the underlying organizational structure. In turn, the assessments lead to some reflections in which way decisions as to an efficient structure can be taken.

II. Objectives of Decentralized Planning Units

Successful decentralization of responsibility from general management to local planning units has to take care of the overall cost of production which is to be minimized while taking general considerations as e.g. the given production tasks or machine capacities into account. All planning units in the production area, especially those units in charge of arranging the production process, will be confronted with this objective and will consequently have to pursue a thoroughgoing objective development for their own area of responsibility.

To be able to assess better the way in which such conclusions are reached and the decision alternatives those responsibles should have, it is essential to take notice of the basic organizational design. If in a general depiction the organizational structures can be labelled input- and output-orientated, then the reference to production planning allows of an interpretation which goes further. Outputs are, according to the type of production, the finished products or jobs, whereas for the present purpose, the basic inputs are the means of production, especially the utilized machines. Thus two types of structure present themselves: on the one hand a product-based line organization and on the other hand a machine-based process organization. In a line organization machines of each type are assigned to a local planning unit so far as the product line requires those machines. On the other hand, a process organization is characterized by an assignment of all machines of the same type to one specific planning unit (see e.g. AXSÄTER 1982). Both structures are illustrated in Fig. 1.1 by means of a simple example for which the differentiation is practical and evident. The figure shows a two-stage two-product production. For each stage the capacity is defined by two identical machines. The respective planning units will be known as divisional management in the case of a line organization or as manufacturing management when a process organization is prefixed. It is now first of all of interest to see how the units can adopt the overall cost objective with respect to their limited planning responsibility.

Let us examine the situation of a divisional manager who is responsible for a certain product line within the framework of a line organization. His main

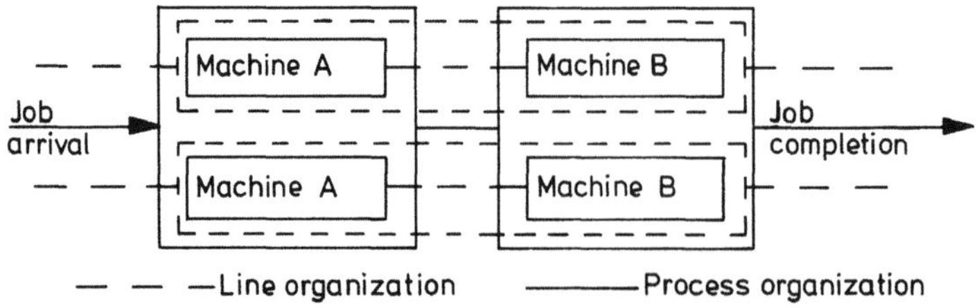

Fig. 1.1. Different organizational structures of a two-stage two-product production process

priority will be to deliver the products as speedily as possible after receiving the orders so that the cost of capital, which is tied up during production, can be reduced. Because of this, the main distribution of the divisional manager towards a reasonable costing-arrangement lies in the minimization of the capital turnaround or – which is equivalent – in the minimization of the flowtime of those types of product for which he has been made responsible. Any other machine-orientated objective of reducing costs, for instance by loading machines to full capacity, can only be of secondary importance to him, as relatively homogeneous products require a similar technological process and thereby simplify machine scheduling.

On the other hand a manufacturing manager's interest will be focussed on loading the machines for which he is responsible to full capacity, in order to avoid costs resulting from idle times, especially as his type of machine will be making heterogeneous products with varying processing times. As he can only see a single-stage part of the whole production, and as responsibility for further parts, i.e. other machine types, lies in other hands, a minimization of flowtime in this particular unit would not be practical without detailed knowledge of the other machine schedules and without being able to foresee the consequences of job delays.

Basically, the machine- and job-orientated objectives, maximizing the machine load and minimizing the flowtime of the job, are not absolutely compatible. GUTENBERG (1983) described it aptly as the "dilemma of production planning". With the imposition of a definite organizational structure, however, a preliminary decision is made to avoid this dilemma, either in favour of job-orientation or of machine-orientation.

The following example shall demonstrate why exclusive concentration on the flowtime, i.e. job orientation, is not always practical, when a machine-based process organization is established and coordination between the different units as regards the harmonizing of machine schedules is not fine enough. It is to be supposed that operation of jobs takes place in such a way that the three different jobs 1, 2, and 3 have each to be processed on the machines A, B, and C, in this sequence, before they are finished. There are two of each type of machine, so that the machines A1, A2, B1, B2, C1, and C2 can be used. The processing times are set as shown in Table 1.1.

If all jobs are ready at the same time, which can be taken for granted, the manufacturing manager A's decision is unproblematical. With the job sequence (2, 1) on machine A1 and the parallel processing of job 3 on A2 idle times of the machines before job completion are avoided and the mean waiting time of the jobs in front of the machines is minimized. If the manufacturing managers B and

Table 1.1. Processing times

$t_{1A} = 1$	$t_{1B} = 4$	$t_{1C} = 1$	(job 1)
$t_{2A} = 1$	$t_{2B} = 5$	$t_{2C} = 5$	(job 2)
$t_{3A} = 3$	$t_{3B} = 1$	$t_{3C} = 7$	(job 3)
(machine A)	(machine B)	(machine C)	

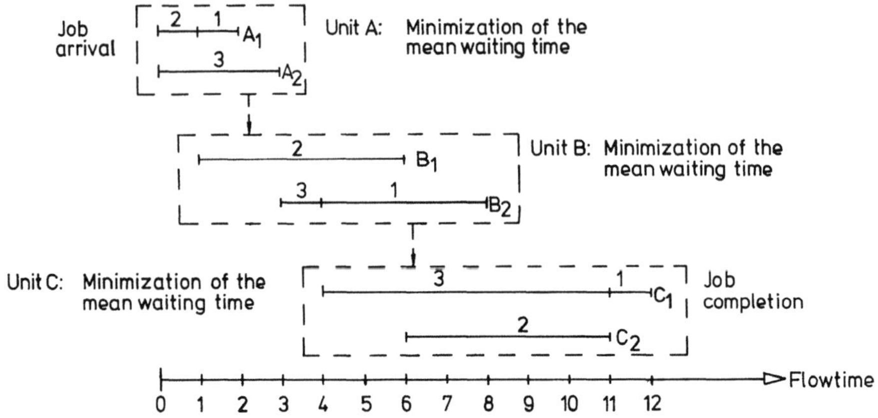

Fig. 1.2a. Process organization (pure job orientation)

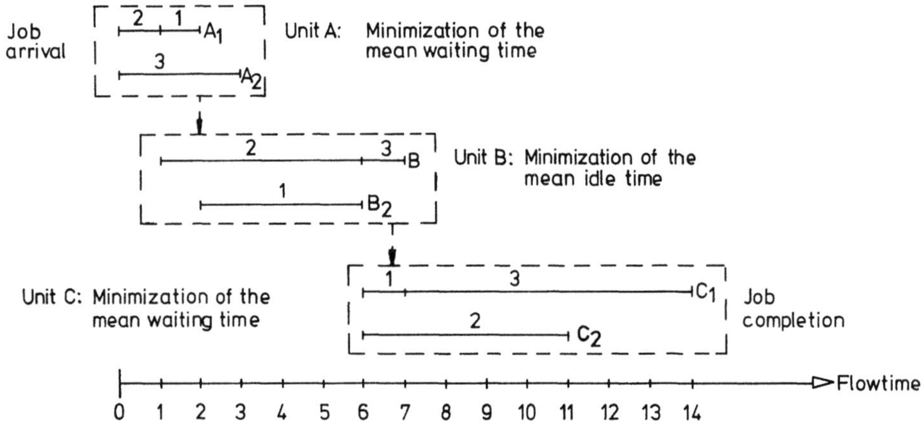

Fig. 1.2b. Process organization (combined job/machine orientation)

C also make decisions based on the waiting times there result machine schedules which are shown in Fig. 1.2a. The mean flowtime of all jobs amounts to $(12+11+11)/3=34/3$ units of time. If manufacturing manager B had, however, tried to reduce idle time by means of the job sequence (2, 3) on B1 and parallel processing of job 1 on B2, this would have caused an immediate increase in mean waiting time for the jobs before the commencement of the operations on the second stage. Nevertheless, the total mean flowtime for all jobs would have been lower, namely $(7+11+14)/3=32/3$ units of time, although job orientation had not existed at all stages of the procedure (see Fig. 1.2b).

The reason for the relatively late completion time of jobs in spite of general job orientation is to be found in the limited flow shops for which the manufacturing managers take responsibility. Even if the jobs manage to leave a processing stage as quickly as possible the exit sequence of the jobs plus an insufficient knowledge of the following machine schedules can determine a poor flowtime.

It is to be noted that a coordination mechanism becomes necessary in the example, whereby the preliminary stage passes on exact information about its machine schedules, so that preparations can be made at the next stage, e.g. concerning the setup of machines. If this were not the case, and information were to arrive only in the form of partly-finished products, then there would be much greater delay. Nevertheless, coordination between the planning units does not go so far as to include an agreement as to the overall job sequences.

III. The Choice of a Sequencing System in Dependence on the Organizational Structure

The balance between structure and objectives must be differentiated when additional coordination measures are established. For example, in a situation of underemployment, the then existing idle times of machines can hardly be treated as part of costs, for instance in the form of opportunity costs or lost profits, so that the "full capacity" objective becomes irrelevant, regardless of which organizational structure has been chosen. On the other hand, a fully employed firm might not pay too much attention to the flowtime of their products if its customers have taken tardiness into account anyway. I.e. whenever such a development in employment is noticed the customers cause external priorities, in that their behaviour is closely connected with the objectives of sequencing. Hence, to guarantee an efficiently decentralized production process with regard to the fulfilment of prefixed objectives, a production sequencing system has to be chosen which includes special coordinative elements. Meanwhile a great number of priority rules are put into practice (see e.g. DANNENBRING 1977) with which, on the basis of simple characteristics, such as arrival time or required job-processing time, decisions are taken about the sequence in which the jobs are to be routed over the machines. It is precisely this instrument for the determining of priorities which shall be now dealt with.

As the problem of defining objectives is no longer to be the subject of analysis here, in view of what has already been said and in order to be able to leave such arguments out in the following, it is to be assumed that the firm in question has sufficient capacity to cope with all jobs, and that the planning units are in agreement that all jobs are to be finished as rapidly as possible. The transformation of time into the corresponding costs caused by the utilization per unit time of the means of production, material and labour is not always clear-cut and in this case it is not really helpful. AXSÄTER (1982) shows a possible way of including cost when dealing with forms of coordination in the production process in that he attempts to express the time taken for coordination and information in terms of capital tied up during production.

As already mentioned, the sequencing problem here to be discussed is enlarged by a problem of coordination and, based on that, cooperation which lies in developing a fast and reliable flow of information between units. The question arises in particular as to how and when information about individual job processing is to be passed on to the next stage in order to bring out a favourable job schedule. To answer this, let us return to the simple case presented in Fig. 1.1. Let it be assumed that both types of job reoccur frequently and that their arrival times are representable in a queueing system with POISSON-input (see SAATY 1961, p. 33) in which the rate of arrival is λ, which means that on average λ jobs arrive per unit of time, λ_1 out of them being jobs of type 1 and λ_2 being jobs of type 2. The expected service rates, namely the average numbers of processings per unit of time, may also presupposed. Depending on the type of job, 1 or 2, and the type of machine, A or B, they are μ_{1A}, μ_{1B}, μ_{2A}, and μ_{2B} whereby the individual processing times are distributed exponentially. In addition, sufficient capacity is present, so that the service rates will be higher than the arrival rates. A positive probability of machine failure during processing intervals is not taken into account.

In order to give details of the quality of an organizational structure regarding the expected flowtimes required for jobs, two different priority rules will be taken in this dynamic sequencing model as a basis for conveying fundamental knowledge about what happens in the individual units. Furthermore, the information necessary for a reduction in flowtime will be discussed.

(a) "First Come, First Served"-Service Discipline

One natural service-discipline which is very easy to put into practice because there are no special coordination efforts is the "first come, first served" (FCFS)-principle:

- Where the organization is on product lines, further information from outside the unit is unnecessary as the whole processing procedure for a job is in one hand. An exchange of information between the two units would, at the most, take place only to report the existence of free capacity which could be used by the other unit. On the grounds that it is assumed that capacity is sufficient this argument will not be further pursued.
- If the organization of the production process is machine-orientated, i.e. if each unit consists of only one type of machine, then it is unnecessary to inform the second stage about progress made in the first stage. If there are jobs already waiting for processing by machine-type B – and this is more than likely when λ_i converges towards μ_{iB}, $i = 1, 2$, – then the job waiting the longest is the only one of interest to unit B with regard to further processing, no matter what is being done in the first unit A.

Within a product-based line organization, the expected waiting times $E(W^L_{ij})$ for the individual types of jobs $i = 1, 2$ in front of the machines $j = \text{A, B}$ are given by

$$E(W^L_{iA}) = \frac{\varrho_{iA}}{\mu_{iA}(1-\varrho_{iA})}\,, \qquad i = 1, 2,$$

and

$$E(W^L_{iB}) = \frac{\varrho_{iB}}{\mu_{iB}(1-\varrho_{iB})}\,, \qquad i = 1, 2$$

(see for this standard result e.g. HILLIER and LIEBERMAN 1974, p. 399).

$$\varrho_{iA} = \lambda_i/\mu_{iA}$$

and

$$\varrho_{iB} = \lambda_i/\mu_{iB} \ ,$$

respectively, denote the utilization factors for the machines or so called "traffic densities". It should further be noted that the exit rate of jobs from machine A is the same as their arrival rate (BURKE 1956) and at the same time is, obviously, the arrival rate in front of machine B. Add the average processing times $1/\mu_{ij}$, then the expected rates for the time spent at each individual stage are obtained as follows:

$$E(V_{iA}^L) = \frac{1}{\mu_{iA}(1-\varrho_{iA})} \ , \quad i = 1, 2,$$

and

$$E(V_{iB}^L) = \frac{1}{\mu_{iB}(1-\varrho_{iB})} \ , \quad i = 1, 2 \ .$$

The expected flowtimes $E(F_i^L)$ of the jobs – if transportation times can be neglected – are then

$$E(F_i^L) = E(V_{iA}^L) + E(V_{iB}^L) \ , \quad i = 1, 2 \ .$$

If, on the other hand, the FCFS-discipline is used within a machine-based process organization then this can be described by a queueing system with two parallel service stations on each stage, whereby no difference is made between the two types of job 1 and 2, jobs of both types being put into a common queue. Waiting times can now be determined, independent of the type of job, as follows:

$$E(W_{iA}^P) = \frac{\varrho_A^2}{\mu_A(2-\varrho_A)(2+\varrho_A)} \ , \quad i = 1, 2 \ ,$$

and

$$E(W_{iB}^P) = \frac{\varrho_B^2}{\mu_B(2-\varrho_B)(2+\varrho_B)} \ , \quad i = 1, 2 \ ,$$

(see e.g. NEUMANN 1977, p. 398). The average processing times on the machines as well as the utilization factors are defined by

$$1/\mu_j = \lambda_1/\lambda \cdot 1/\mu_{1j} + \lambda_2/\lambda \cdot 1/\mu_{2j} \ , \quad j = A, B \ ,$$

and

$$\varrho_j = \lambda/\mu_j \ , \quad j = A, B \ .$$

As before the expected flowtimes are calculated as

$$E(F_i^P) = E(W_{iA}^P) + 1/\mu_{iA} + E(W_{iB}^P) + 1/\mu_{iB} \ , \quad i = 1, 2 \ .$$

The expected values $E(F_i^L)$ and $E(F_i^P)$ can now be taken for a comparison between a line organization and a process organization. When all the aforesaid requirements are met, then it can be shown in a simple calculation that the inequalities

$$\lambda_1/\lambda \cdot E(W_{1j}^L) + \lambda_2/\lambda \cdot E(W_{2j}^L) > E(W_{1j}^P) = E(W_{2j}^P) \ , \quad j = A, B$$

always hold. Hence

$$\sum_{i=1}^{2} \lambda_i/\lambda \cdot E(F_i^L) > \sum_{i=1}^{2} \lambda_i/\lambda \cdot E(F_i^P)$$

follows immediately so that, minimizing mean flowtime, process organization becomes superior. As long as the FCFS-discipline is valid line organization cannot dominate, because the possibility of adjusting jobs among machines of one type is dispensed with and there are no other positive coordination effects. Line organization remains inferior even when dealing with jobs with identical arrival and service parameters.

(b) "Shortest Processing Time"-Service Discipline

By reason of a "first come, first served"-discipline the problem of the supply of information between organizational units did not have to be considered. It is well known, however, that the FCFS-principle leads to comparatively long flowtimes, regardless of the organizational structure (BAKER 1974, p. 214, and many others). Especially promising, on the other hand, among the great number of applied priority rules (for a listing see e.g. PANWALKAR and ISKANDER 1977), is the choice of job processing according to the processing time (BAKER 1974, p. 18). The "shortest processing time" (SPT)-rule gives precedence to those jobs which would appear to have the shortest imminent processing times. It is, of course, disadvantageous for those jobs which require difficult operations, but, on the whole, does lead to a comparatively low mean flowtime.

If the SPT-discipline is included in the criteria for an assessment of the organizational structure a further problem arises which has to be taken into account. The underlying question is, from which set of jobs is the one with the shortest processing time to be taken and so to be dealt with next? It seems obvious that this problem should not then be solved when the relevant machine is free for processing. On the contrary, preparations will have to be made, for example as to the allocation of material and labour for a specific job or as to the setup of a machine, so that the processing begin is not delayed unnecessarily. At this anticipatory planning level not all jobs have already been through the earlier processing stages, which would be available for selection when the machine is free, so that sufficiently exact information is of importance in order to be able to include all these jobs as well in the planning. This situation is depicted in Fig. 1.3. The time period $a = t_2 - t_1$ defines how early information about the anticipated completion date t_2 of job i on machine A is sent to the unit B for further planning. Accordingly, $a = 0$ means that early information is dispensed with, so that planning for further processing of job i can only commence when the job has been finished on machine A and arrives in front of machine B. On the other hand time period b illustrates the planning support necessary when the machine B is free again at t_3. Flowtime of job i is expected to be delayed when

$$b - a > 0 \ .$$

Even if job i is of utmost priority one will have to take a delay of this length into account. Generally speaking, the extent of a minimum delay caused by unpunctual information can be expressed as

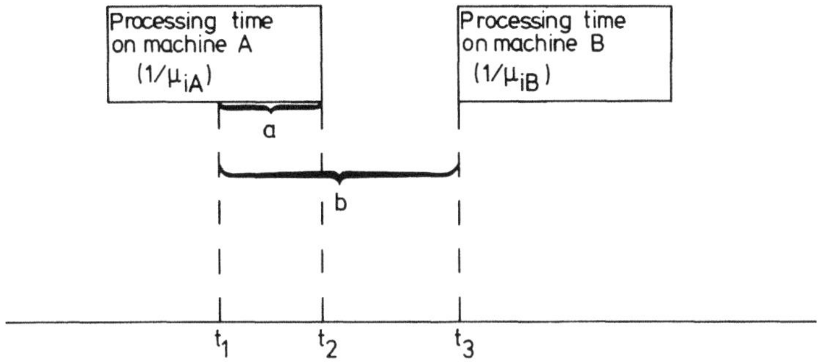

Fig. 1.3. Information lead for planning support in multistage production

$$c = \max(0, b-a) \ ,$$

whereby a late information flow as well as a strong desire for planning support strengthen this delay. Assuming that a, b and therefore c depend on the organizational structure, but are identical for all job-types in that they represent identical expected values of corresponding random variables, then their independence from the other random variables is easily verified.

Let the sequencing problem now be modified in such a way that the service discipline on machine B is determined by applying the SPT-rule, whereby job 1 is always given precedence over job 2 because of its shorter average processing time. Within one priority class the FCFS-discipline is still valid. To simplify matters, the first processing stage continues operating on the FCFS-principle, because the internal information set-up is of no importance there anyway. Thus, only the waiting times in front of machine B must be calculated anew. As FERSCHL (1964a, p. 159) deduces, the expected values of those waiting times can be derived as follows in the case of a process organization:

$$E(W_{1B}^{P'}) = \frac{2E(W')}{2-\varrho_{1B}}$$

and

$$E(W_{2B}^{P'}) = \frac{4E(W')}{(2-\varrho_{1B})(2-\varrho_{B})} \ .$$

The random variable W' symbolizes that part of the waiting time of any job which is caused by other jobs still being processed, and which have to be finished, when the job joins the queue. Let the random variable R describe the remaining service time of any job on arrival of another job in front of the same machine, and let p_b be the probability of the machines of type B being in use, then

$$E(W') = p_b E(R) + (1-p_b) \cdot 0 = p_b E(R)$$

follows. Because of

$$E(R) \quad = \frac{1}{2\mu_B}$$

(see FERSCHL 1964a, p. 52) and

$$p_b = \frac{\varrho_B^2}{2+\varrho_B}$$

(see NEUMANN 1977, p. 398) one obtains directly:

$$E(W_{1B}^{P'}) = \frac{\varrho_B^2}{\mu_B(2-\varrho_{1B})(2+\varrho_B)}$$

and

$$E(W_{2B}^{P'}) = \frac{2\varrho_B^2}{\mu_B(2-\varrho_{1B})(2-\varrho_B)(2+\varrho_B)} \ .$$

Finally, these results can be used to compute the expected flowtimes

$$E(F_i^{P'}) = E(W_{iA}^{P'}) + 1/\mu_{iA} + c + E(W_{iB}^{P'}) + 1/\mu_{iB} \ , \quad i = 1,2 \ .$$

Whereas a firm whose planning units are organized according to the production process still possesses a genuine option as regards the sequencing of jobs, this is not true for a product-based line organization. Because of the similarities between the jobs within a special product line, there hardly exist different priority classes with respect to the processing times, so that the FCFS-discipline or some other "natural" priority rule will come to the fore. For example, the "last come, first serviced"-method can be used as such a service discipline when, say, preprocessed metal sheets are stacked on top of one another on arrival and the pile is then worked on from the top to bottom. Waiting times and flowtimes still correspond to those of the FCFS-principle (see FERSCHL 1964b).

Thus, the SPT-priority rule will mainly be at disposal in the case of a process organization. First of all, let us therefore regard the consequences which arise for a process organization when substituting the FCFS-rule by the SPT-discipline. As expected, the inequalities

$$E(W_{1B}^{P}) > E(W_{1B}^{P'})$$

and

$$E(W_{2B}^{P}) < E(W_{2B}^{P'})$$

hold, because of

$$\varrho_{1B} < \varrho_B < 1$$

and because the partition of jobs into two priority classes places job 1 at an advantage and accordingly job 2 at a disadvantage. The inequalities

$$\lambda_1 E(W_{1B}^{P}) + \lambda_2 E(W_{2B}^{P}) > \lambda_1 E(W_{1B}^{P'}) + \lambda_2 E(W_{2B}^{P'})$$

is also easily estimated and demonstrates that the service discipline based on the shortest processing time dominates the FCFS-discipline, when the time lag c, caused by unpunctual information, is ignored. If, however, c is taken into consideration in sequencing according to the SPT-discipline, then this relationship must be questionned. The FCFS-discipline remains only then inferior with respect to the mean flowtime when

$$c < \frac{(\lambda_1\varrho_B - \lambda_{1B})\varrho_B^3}{\lambda^2(2-\varrho_B)(2+\varrho_B)(2-\varrho_{1B})}$$

is satisfied. If, for example, the arrival rates of both types of jobs differ strongly then the efficiency of the SPT-rule is just as considerably reduced as in the case of a rather low utilization factor ϱ_B. As is further evident this service discipline needs also no longer be superior when the mean processing times of both types of job are approximately equivalent. Hence, c must be kept as short as possible, as otherwise the choice of priorities on the FCFS-service discipline would dominate. In turn, a longer delay c is only then in accordance with the SPT-discipline when traffic-density is generally high, processing times vary widely and the different types of job can be considered in roughly the same frequency.

These last reflections were all based on a comparison of two priority rules only. Together with the examination of the process and line organizations with regard to the FCFS-service discipline they support a general assessment of both organizational structures as far as effects on the production planning system are concerned. The main result which can be derived is that, with the assumptions given, SPT-sequencing should be applied if a process organization is established and if planning support by informational means is sufficient. Here, the SPT-rule performs better than a FCFS-service discipline. In turn, a sequencing system based on the FCFS-rule becomes efficient in the case of a line organization or if the planning lag caused by informational delay cannot be reduced by no means. Assessing the organizational structures here in question, it emerges from the discussion that line organization cannot dominate process organization, unless an inappropriate priority rule is chosen. Nevertheless, inefficiency of a line organization may turn into efficiency as soon as product lines consist of several jobs with obviously different processing times (see e.g. AXSÄTER 1982).

IV. Further Extensions

The fact that two priority rules and two organizational structures have been dealt with does not by any means signify that an entrepreneurial decision can be taken only by using these alternatives. Other methods of realizing the organization of production process are always possible. The options for solutions put forward here have simply the advantage that they make extreme positions clear. While one priority rule demands hardly any expenditure on coordination or information, but then gives less good results, the requirement for planning and information with the other rule is considerable, though this can be of benefit in minimizing the mean flowtime of jobs. It is of course conceivable that some sort of compromise can be found, deviating from these rules, which takes the conflicting criteria of the effort to provide information and the quality of the results into consideration in another form.

On the other hand, further expansions of the effort towards coordination, beyond merely informing the next process stage in advance as described, would lead, as a final consequence, to an exact solution of sequencing problems within the framework of a centralized system. The consequent effects on the production flow are sufficiently well known (see e.g. BAKER 1974). A centralized production management given, the choice of the organizational structure would first of all be of importance under the aspect of the rapid and reliable availability of information to the central authority. The negligible practical relevance of the centralization of planning and decision-making for routine problems, such as usually have to be managed during the production flow, is dwarfed by the extraordinary effort needed to supply that information and by the comparatively insignificant improvement in flowtimes or in machine loads.

Furthermore, the discussion has generally differentiated between a product-based line organization and a machine-orientated process organization. This was done in order to restrict the subject under discussion to its essential criteria, which are limited by its objectives. A more detailed examination, on the other hand, would show that each of these structures is able to contain varying forms of organizational design. In the case of line organization the most important examples, and certainly the ones which are the most often encountered in practice, are assembly lines and fabrication lines. The additional problems which are linked to this kind of extended differentiation, such as the nature of the jobs, the facilities for intermediate storage or the setting-up of a suitable transport system, do of course affect the flowtimes of the jobs and thereby the choice of the organizational structure.

V. Conclusions

The growth of a firm leads, in general, to an intensification of organizational reflections concerning the decentralization of decision-making in the different departments. If a close look is taken at the area of production, it can be easily seen that the choice of an organizational structure is governed by the production objectives and by the attitude of the planning units to the important coordinating factor of information. Basically, a capacity-orientated objective would seem to be more easily attainable by means of a performance-orientated organizational structure, while a job-orientated objective is more favourably influenced by product-orientated structure. In situations in which it is more efficient to leave the objective of full capacity load in the background, a process organization has nevertheless a legitimate claim especially if the means of coordination are so distinct that any job tardiness caused by insufficient information and harmonizing does not have to be expected, or if coordination on a wide basis can be done by means of a simple, standardized sequencing system. BURBRIDGE (1982) points out that even a reorganization of production area on the lines of a structure which proves to be superior can be carried out quite simply and at short notice, so that sufficient flexibility of the organizational structure in the face of unforeseeable developments regarding the jobs can be guaranteed.

As seems obvious, the presented reflections cannot serve immediately for evaluation of a firm's complex production process, because most assumptions are not met in practice. Thus, only a qualitative advice can be intended as the dependence between organizational decisions and planning procedures in the production area is concerned. Before, however, any final decision with regard to a reorganization of that area or a choice of a production planning system is taken, the entrepreneurial parameters have to be taken into account instead.

References

Axsäter, S.: Decentralized Production Planning and Choice of Organizational Structure, in: International Journal of Production Research, 1982, pp. 17–26.

Baker, K.R.: Introduction to Sequencing and Scheduling, New York-London-Sydney-Toronto 1974.

Burbridge, J.L.: The Simplification of Material Flow Systems, in: International Journal of Production Research, 1982, pp. 339–347.

Burke, P.J.: The Output of a Queuing System, in: Operations Research, 1956, pp. 699–704.

Dannenbring, D.G.: An Evaluation of Flow Shop Sequencing Heuristics, in: Management Science, 1977, pp. 1174–1182.

Ferschl, F.: Zufallsabhängige Wirtschaftsprozesse, Wien-Würzburg 1964a.

Ferschl, F.: Auswahlordnungen und die Wirksamkeit von Wartesystemen, in: Zeitschrift für Operations Research, 1964b, pp. 149–160.

Gutenberg, E.: Grundlagen der Betriebswirtschaftslehre, Volume I: Die Produktion, 24th edition, Berlin-Heidelberg-New York 1983.

Hax, A.C., and Majluf, N.S.: Organizational Design: A Survey and an Approach, in: Operations Research, 1981, pp. 417–447.

Hillier, F.S., and Lieberman, G.J.: Operations Research, 2nd edition, San Francisco 1974.

Mintzberg, H.: The Structuring of Organizations, Englewood Cliffs 1979.

Neumann, K.: Operations Research Verfahren, Volume II, München-Wien 1977.

Panwalkar, S.S., and Iskander, W.: A Survey of Scheduling Rules, in: Operations Research, 1977, pp. 45–61.

Saaty, T.L.: Elements of Queueing Theory, New York-Toronto-London 1961.

Chapter 2

Organization and Planning of Vehicle Utilization in a Chemical Firm

GÜNTER FANDEL and JOACHIM REESE

I. Introductory Remarks

Within the framework of production planning the internal transport of goods, i.e. the transport between various locations within a work's grounds, is attributed an auxiliary function which is becoming increasingly important on account of the growing spatial expansion of firms and because the production schedule is otherwise planned down to the last detail. In the forefront of these reflections is the task of planning the utilization of vehicles, in accordance with which – the transport demand and the number of available vehicles supposed as given – all orders must always be carried out efficiently. In order to efficiently accomplish all orders they should be carried out in such a way that all the routes and travel times are kept to a minimum and so that individual operations of the production process are not hindered by unpunctual execution.

Although in the meantime numerous suggestions have been made in the literature as to how this problem can be solved (see e.g. BODIN et al. 1983 or CARRARESI and GALLO 1984) someone trying to practically apply them in a firm often comes up against considerable difficulties because important methodological assumptions cannot in reality be fulfilled. Thus for example it is of particular consequence that, contrary to most theoretical solutions, in reality the extent of the transport demand is not foreseeable at the beginning of the planning process as a large part of the orders are placed during the course of the planning period. In the following, using the example of a medium sized chemical firm – in which the vehicle scheduling has up to now be carried out solely on a "first come, first served" (FCFS)-basis –, it will be shown that it is in fact possible, indeed with a considerable reduction of both time and costs, to carry out a sensible planning on the basis of less intuitive solution methods. The firm examined here which has a turnover of 400 million DM employs 1600 employees and mainly produces chemical specialities. All internal transportation is directed by the transport department which employs fork-lift trucks and electric trucks. Various sorts of orders have been considered here including the transport of raw materials and of semi-finished and finished products between locations in the firm as well as within one of these locations in the firm. The data concerning the orders were collected over a period of one working week, i.e. five days.

II. Description of the Problem

Within the total work's ground of about 270,000 m^2 there are 73 plant locations which can take advantage of the vehicles based in the depot. Of these 73 plant locations only the 37 which regularly require transportation services on a large scale are relevant for the problem to be dealt with here. The relative position of these plant locations to each other and to the vehicle depot has been sketched in Fig. 2.1. In addition, Table 2.1 contains the distances between the various locations. The number of orders made by the plant locations within one day varied between 30 and 47 during the period in which the data were collected. These orders along with the estimated duration of its execution are constantly reported to the management in the vehicle utilization headquarters along with the earliest possible time at which an order can be carried out. The order durations comprise the total period of execution up to the return of the vehicle to the plant location. All the corresponding data are noted on a time registration form. Although the total vehicle fleet consists of 15 fork-lift trucks, on account of outages caused by servicing and repairs or by loans, during the period of investigation regularly only 10 fork-lift trucks, which have the same basic equipment and are able to carry out any orders, were unreservedly available. Other fork-lift trucks are only available at times. The working time of the drivers amounts to 8 1/2 hours per day, including half an hour break, so that each fork-lift truck can be used for a maximum of 480 minutes per day. The speed of the fork-lift trucks within the work's grounds is around 6 km per hour, i. e. 100 m per minute. Thus, the routes

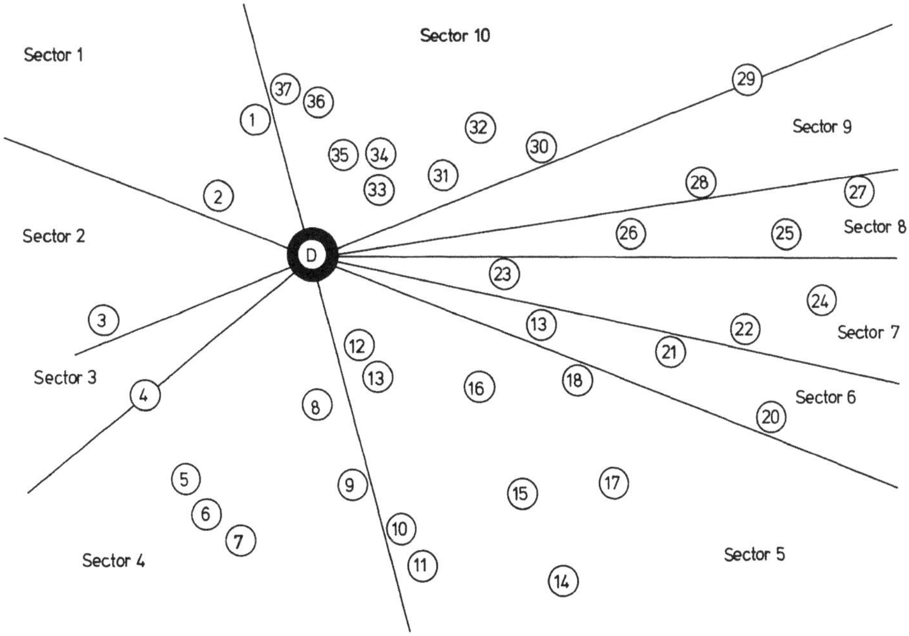

Fig. 2.1. Positions of the plant locations and of the depot

Table 2.1. Distances between the plant locations and the depot (100 m)

	D	1	2	3	4	5	6	7	8	9	10	11	12	13	14	15	16	17	18	19	20	21	22	23	24	25	26	27	28	29	30	31	32	33	34	35	36	37
D		0.9	0.9	1.5	0.9	1.1	1.2	1.5	0.8	0.9	1.5	1.8	0.3	0.6	2.7	2.1	1.2	3.0	2.1	1.4	5.1	3.0	4.8	1.2	5.7	5.1	2.6	6.0	3.3	4.5	2.1	0.9	1.4	0.5	0.6	0.5	0.9	0.9
1			0.3	1.2	0.9	1.5	1.8	2.1	0.9	1.5	2.1	2.7	1.1	1.2	3.3	2.7	2.0	3.6	2.7	2.1	6.0	3.6	5.4	1.8	6.3	5.7	3.0	6.6	3.9	5.1	2.7	1.5	1.8	0.9	1.1	0.6	0	0
2				0.9	0.8	1.2	1.8	2.1	0.9	1.5	2.1	2.7	1.2	1.5	3.3	2.7	2.1	3.9	3.0	2.4	6.0	3.9	5.7	2.1	6.3	6.0	3.3	6.9	4.2	5.4	2.8	1.8	2.1	1.1	1.2	0.9	0.3	0.3
3					0.6	1.0	1.5	1.8	1.5	1.6	2.1	2.5	1.6	1.9	3.6	3.3	2.5	4.2	3.3	2.7	6.4	4.3	6.1	2.5	6.9	6.6	3.9	7.5	4.6	5.8	3.6	2.4	2.8	1.8	1.8	1.8	1.2	1.2
4						0.6	0.9	1.3	0.9	1.0	1.5	1.9	1.0	1.3	3.0	2.7	1.9	3.6	2.7	2.1	5.8	3.7	5.4	1.9	6.3	5.8	3.3	6.7	4.0	5.2	3.0	1.8	2.2	1.3	1.3	1.3	0.9	0.9
5							0.3	0.8	0.9	0.8	0.9	1.5	1.0	1.2	2.5	2.1	1.5	3.1	2.4	1.8	5.4	3.3	5.1	1.8	6.0	5.7	3.0	6.4	3.9	5.1	2.8	1.8	2.2	1.5	1.5	1.5	1.5	1.5
6								0.5	1.1	0.8	0.8	1.0	1.2	1.2	2.4	1.9	1.5	3.0	2.2	1.9	5.2	3.3	4.9	1.8	5.8	5.5	3.0	6.4	3.7	4.9	2.8	1.8	2.4	1.8	1.8	1.8	1.8	1.8
7									1.2	0.8	0.3	0.6	1.3	1.2	1.9	1.6	1.3	2.5	1.9	1.8	4.9	3.0	4.6	1.8	5.4	5.2	2.7	6.0	3.4	4.5	2.5	1.8	2.2	1.9	1.9	1.9	2.1	2.1
8										0.6	1.2	1.6	0.3	0.6	2.4	1.9	1.2	3.0	1.9	1.2	5.1	3.0	4.8	1.2	5.7	5.1	2.5	6.0	3.3	4.5	2.1	0.9	1.5	0.6	0.8	0.8	0.9	0.9
9											0.8	1.0	0.6	0.6	1.9	1.5	0.9	2.5	1.6	1.2	4.8	2.7	4.5	1.2	5.2	4.9	2.4	5.8	3.1	4.3	2.1	1.2	1.6	1.2	1.2	1.2	1.5	1.5
10												0.5	1.3	1.2	1.6	1.3	1.2	2.2	1.6	1.5	4.6	2.7	4.3	1.5	5.2	4.8	2.4	5.7	3.1	4.2	2.2	1.8	2.1	1.8	1.8	1.8	2.1	2.1
11													1.6	1.5	1.3	1.0	1.2	1.9	1.5	1.5	4.3	2.4	4.0	1.6	4.8	4.6	2.4	5.4	3.0	4.0	2.2	1.8	2.1	2.1	2.1	2.1	2.7	2.7
12														0.3	2.4	1.8	0.9	2.7	1.8	1.0	4.8	2.7	4.5	0.9	5.2	4.9	2.2	5.7	3.0	4.2	1.8	0.8	1.2	0.5	0.5	0.5	1.1	1.1
13															2.1	1.5	0.6	2.4	1.5	0.8	4.6	2.4	4.2	0.6	4.9	4.6	1.9	5.5	2.8	4.0	1.6	0.6	1.0	0.6	0.6	0.6	1.2	1.2
14																0.6	1.5	0.8	0.9	1.5	3.0	1.5	2.8	1.8	3.7	3.4	1.6	4.2	2.1	3.0	1.8	2.1	2.2	2.7	2.7	2.7	3.3	3.3
15																	0.9	0.9	0.6	1.0	3.3	1.3	3.0	1.3	3.7	3.6	1.5	4.3	1.9	3.0	1.5	1.5	1.8	2.1	2.1	2.1	2.7	2.7
16																		1.8	0.9	0.5	3.9	1.8	3.6	0.6	4.5	4.2	1.5	4.9	2.2	3.4	1.2	0.9	1.0	1.2	1.2	1.2	2.0	2.0
17																			1.0	1.6	2.4	0.9	2.1	2.1	3.0	2.7	1.3	3.6	1.5	2.2	1.6	2.4	2.2	3.0	3.0	3.0	3.6	3.6
18																				0.8	3.1	1.0	2.7	1.0	3.6	3.3	0.9	4.2	1.5	2.5	0.9	1.3	1.3	1.9	1.9	1.9	2.7	2.7
19																					3.7	1.6	3.3	0.3	4.2	3.9	1.2	4.6	1.9	3.4	0.9	0.6	0.8	1.2	1.2	1.2	2.1	2.1
20																						2.1	0.8	4.2	0.8	0.8	2.7	1.3	2.1	1.2	3.3	4.5	4.2	5.1	5.1	5.1	6.0	6.0
21																							1.8	1.8	2.5	2.2	0.8	3.1	0.6	1.6	1.2	2.1	1.9	2.7	2.7	2.7	3.6	3.6
22																								3.6	0.9	0.6	2.2	1.3	1.5	0.6	2.7	3.9	3.6	4.5	4.5	4.5	5.4	5.4
23																									4.5	4.2	1.3	4.8	2.1	3.3	1.0	0.3	0.6	0.9	0.8	1.0	1.8	1.8
24																										0.5	3.1	0.6	2.2	1.2	3.4	4.8	4.5	5.4	5.4	5.4	6.3	6.3
25																											2.7	0.9	1.9	0.8	3.1	4.3	4.0	4.9	4.9	4.9	5.7	5.7
26																												3.6	0.8	1.9	0.5	1.6	1.3	2.2	2.2	2.2	3.0	3.0
27																													2.8	1.5	4.0	5.1	4.8	5.7	5.7	5.7	6.6	6.6
28																														1.2	1.2	2.4	2.1	3.0	3.0	3.0	3.9	3.9
29																															2.4	3.6	3.3	4.2	4.2	4.2	5.1	5.1
30																																1.2	0.9	1.8	1.8	1.8	2.7	2.7
31																																	0.5	0.6	0.5	0.8	1.5	1.5
32																																		0.9	0.9	0.9	1.8	1.8
33																																			0	0	0.9	0.9
34																																				0	1.1	1.1
35																																					0.6	0.6
36																																						0
37																																						

reproduced in Table 2.1 can at the same time be interpreted as travel times per minute. The existing route network is symmetrical, i.e. there are no one-way streets. All vehicles always begin and end their daily tour in the depot.

The task of the vehicle utilization planning now consists of carrying out all the orders recorded for a particular day on the same day whilst at the same moment keeping the travels between two orders and the waiting times to a minimum. At the beginning of the daily employment of the fork-lift trucks the orders have not yet all come in and some are reported during the course of the working day. In addition, it has to be taken into consideration that the loading or unloading of a lorry is always to be given priority and carried out as soon as another vehicle is free again. It is acceptable for an order to be shared by several fork-lift trucks but this practice should be limited to large orders which anyway could not have been managed by one fork-lift truck.

III. Actual Survey of Vehicle Utilization

The procedure adopted upto now in the transport headquarters for the control of vehicle utilization was such that in respect of the restrictions mentioned the orders were divided between the available vehicles, where necessary in accordance with the FCFS principle. If several vehicles were available at the desired time then it was up to the managing clerk to decide upon the final allocation with a view to minimizing the travel time to the plant location in question. If more orders have come in than there were vehicles available then a waiting list was drawn up. Obviously, with this procedure of vehicle employment the number of tours planned and those carried out was subject to daily fluctuations. The results of the investigation period are collated in Table 2.2.

Table 2.2. Results of the actual survey

	1st day	2nd day	3rd day	4th day	5th day	Σ	Ø
Number of orders	43	47	39	39	30	198	39.6
Number of tours	11	11	11	10	8	51	10.2
Total travel times (minutes)	157.4	164.9	113.8	121.2	83.3	640.6	128.1

The sum of the travel times is relatively high here because the location in which a vehicle is used is solely determined by the sequence in which the orders come in with the result that long trips must be undertaken through the work's grounds. For this reason with the same data a solution method was tried out, parallel to the existing one, which without special conditions could be realized by the firm and which is based on a prestructured working area for the individual vehicles.

IV. A Heuristic Method of Solving the Vehicle Utilization Problem

The deciding factor for the choice of the solution method to be explained in the following was the conception of a method which on the one hand can fall back on a suitable familiar procedure but which on the other hand takes the conditions of the firm into account. Both on account of the dimension of the problem as well as with a view to ensuring a flexible implementation, so as to facilitate quick decisions without the use of a computer, in the selection of a method exact procedures were excluded right from the start. With regard to the desired structuring of the working area with corresponding allocation of the vehicles it seemed appropriate to test a heuristic method on the basis of the "cluster first, route second" (CFRS)-technique (see e.g. BODIN and GOLDEN 1981 as well as DOMSCHKE 1982). Three problem areas were observed here:

- Determination of the number of tours (capacity problem)
- Allocation of the orders to the tours (assignment problem)
- Laying down of routes (sequencing problem).

In order to deal with the first two problems it is expedient, taking the planned average demands for transport services from the individual plant locations into consideration, to firstly establish a relatively stable initial solution. This should consist of a planned number of tours and a planned sector organization with regard to the utilization of the vehicles. The data concerning the planned demands for transport services stem from the experiences of the plant locations and from average values gathered over a longer period. Both problems can be dealt with in a one step procedure, namely the so called clustering. For this purpose the sweep algorithm from GILLET and MILLER (1974) will be used. It can be described as follows:

The ground plan of the plant locations and of the depot (see Fig. 2.1) is covered with a coordinate system, whereby the depot has to be in the coordinate source. Starting from any one plant location the neighbouring plant locations are then successively determined (with a polar-coordinate angle which becomes constantly greater) by revolving the plane in an anti-clockwise direction; the plant locations are assigned to the first tour or to the first sector as long as the sum of the planned demands for transport services from these locations, including the trips between the neighbouring plant locations as well as there and back to the depot, does not exceed the maximum vehicle utilization time. Then the other planned tours or sectors will be established in the same way. With the completion of this allocation process the number of sectors and of planned tours are ascertained. This number can be estimated down from the minimum number of tours [planned average demands for transport services by all plant locations (minutes/day)/maximum utilization time of a vehicle (minutes)] ([x] means the smallest whole number $\geq x$), which, if the travel times between the plant locations are not observed, would be sufficient to deal with all planned orders.

The initial solution is to serve the purpose of preventing fundamentally new capacity and assignment problems having to be solved every day. Nevertheless, the discrepancy between the actual orders and the long term planned average values has to be taken into account. This occurs with the help of an ABC-analysis by means of which the largest contractors amongst all the plant locations – which use 75% of the time capacity – are to be dealt with first. In addition, these locations are questioned in advance about the orders for the following day so that alterations in the number of tours on account of the discrepancies between the actual and the average orders are foreseeable and a corresponding disposition with regard to the capacity reserve can be carried out in time. Thus, the planned number of tours can be corrected on the day before. If the demand for transport fluctuates to a great extent then it is still possible to undertake corrections on the working day itself.

On account of the incomplete preadvice and precise wishes as to the time of execution the routing problem within the various sectors is of less importance. If the vehicle in question is available at the appointed time then the order can be carried out without delay. If the vehicle is, however, occupied there are two alternatives:

- The reported order is placed on a waiting list. This is always worked according to the procedure of the best successor (see e.g. TYAGI 1968).
- The reported order is handed over to another neighbouring tour for completion. This possibility is considered when orders have to be carried out immediately (loading and unloading of lorries) or when the vehicle responsible is at full stretch. A vehicle is at full stretch when the order concerned cannot be completed on the same day or when an unreasonable waiting-time occurs.

With this approach a choice has been made between the numerous procedure heuristics which are available for the problem to be dealt with here. The most well known of these, besides the sweep algorithm, are the savings method (CLARKE and WRIGHT 1964) as well as the greedy algorithm (see EDMONDS 1971 or FRIEZE 1978) which has been modified in many respects. Although the "worst case"-behaviour has also been analysed for some methods (e.g. FRIEZE 1978 or ONG and MOORE 1984) it is not easy to subject individual procedure forms to a comparative assessment. In the case of the sweep algorithm which has been presented above this is difficult because the underlying problem has been changed by important premises like the acceptability of partial services or the initially uncertain transport demand. In spite of the fact that these structures can hardly be compared with one another some results, like those of PAESSENS (1981), for example, derived from empirical investigations, may reveal something about the merits of the sweep technique when certain criteria are fulfilled. Thus, the application of the sweep technique produces more favourable results than the savings technique,

- the more orders can be dealt with per tour;
- the more central the position of the depot is;
- if the plant locations are fairly equally distributed over the work's grounds.

A comparison of procedures was neither striven for nor carried out here, because the task at hand was clearly to solve the problem and not to evaluate the methods. Nevertheless, with regard to an application of the sweep method, these statements, against a background of concrete problem data, do at least point to favourable results. These are presented and commented upon in detail in the following section.

V. Results

Independent of the investigation carried out, the data concerning the average planned utilization of transport services by all the plant locations are always available because of a regular survey. At the time of the examination this amounted to 4,245 minutes/day (Table 2.3). Thus, it is possible to immediately ascertain the average minimum number of tours/day on account of the existing vehicle utilization restrictions as being [4,245/480] = 9. By adding the average travel times, which have not yet been taken into account, the average planned number of tours can also be established. With the results of the actual survey (Table 2.2) an upper limit is revealed for this number in the case in hand of [(4,245 + 128.1)/480] = 10. This means that a sufficient vehicle capacity can be assumed whereby greater fluctuations in the order applications stand opposite a capacity reserve in the form of the other fork-lift trucks which can only be used occasionally.

The exact planning of the number of tours and the assignment of the plant locations to the individual tours (sectors) is then the object of the clusterings using the sweep procedure. Beginning with plant location 1 sector 1 is formed first followed by the other sectors. The exact result of this allocation is reproduced in Table 2.4. In Fig. 2.1 the sector division is also graphically visualized. It is important to note here that although hypothetical routes are assumed in the various sectors, in order to provide criteria for the necessary travels, these routes cannot, however, be binding for the effective daily planning because there the FCFS principle is always decisive. In accordance with this, less favourable routes which diverge from the hypothetical course occasionally result in the overloading of the vehicle concerned and, thus, lead to a correction of the initial assignment as well as to an above average transport demand. It is furthermore to be observed that the plant locations 4 and 29, with a planned utilization per day of 630 minutes each, cannot be met by one vehicle on its own so that both plant locations must be served by two vehicles. Moreover, the consultation carried out on the previous day with the largest contractors with 75% of the order volume or 3,183 minutes of planned utilization – these plant locations are

Table 2.3. Average planned utilization of transport services (minutes/day)

Plant location	1	2*	3*	4*	5	6	7	8	9	10	11*	12	13
Average planned utilization	45	420	450	630	120	24	12	36	99	54	180	1	45
Plant location	14	15	16	17	18	19*	20	21	22*	23	24	25*	26
Average planned utilization	75	6	75	1	1	183	60	21	420	11	18	270	30
Plant location	27	28	29*	30	31	32	33	34	35	36	37	Σ	∅
Average planned utilization	150	75	630	1	30	1	21	18	2	15	15	4,245	115

* In these plant locations a survey is carried out the previous day.

Table 2.4. Fundamental assignment of the plant locations (sector division) with the help of the sweep procedure

Sector	Assigned plant locations	Hypothetical tours	Total utilization time (minutes)
1	1, 2	D-1-2-D	467.1
2	3	D-3-D	453.0
3	4	D-4-D	480.0
4	4, 5, 6, 7, 8, 9	D-4-5-6-7-8-9-D	437.8
5	10, 11, 12, 13, 14, 15, 16, 17, 18	D-10-11-12-13-14-15-16-17-18-D	450.4
6	19, 20, 21	D-19-20-21-D	274.2
7	22, 23, 24	D-22-23-24-D	467.6
8	25, 26, 27	D-25-26-27-D	449.3
9	28, 29	D-28-29-D	480.0
10	29, 30, 31, 32, 33, 34, 35, 36, 37	D-29-30-31-32-33-34-35-36-37-D	348.0

Table 2.5. Planned and actual order volumes during the period of investigation (in minutes)

	1st day	2nd day	3rd day	4th day	5th day
Result of the survey of planned orders (75% of the order volume, 8 plant locations)	2,700	3,300	3,180	3,240	3,090
Projection of the survey of planned orders for 100% order volume*	3,600	4,400	4,240	4,320	4,120
Actual order volumes	3,660	4,110	4,305	4,155	3,675

* The projection is carried out under the premises that the demand of those plant locations questioned correlates strictly positively with the demand of those not questioned.

ascertained on the basis of Table 2.3 and are marked with a star – reveals during the period of investigation that the capacity reserve probably will not have to be employed. The effective execution times later confirm that this needs also not happen at short notice. This is clearly shown by the data listed in Table 2.5. Additional tours were only necessary in such cases when the total order volume V, on account of $[(V+L)/480] > 10$, exceeds $(4800-L)$ minutes, whereby L symbolizes the travel times. Even with a careful estimation $L = 150$ this prerequisite never occurred during the investigation period. A short term reduction of the number of tours is on the other hand not intended from the start. This would, however, be conceivable if other utilization possibilities existed for the vehicles which have been thus economized upon. The discrepancy between the projected order volume and the actual volume, which on some days falls considerably below the projection, is mainly attributable to occasional order cancellations rather than to inaccurate estimates of execution times.

The results of the effective specifications for the tours are recorded in Table 2.6. Here, it is noticeable that each day between three and five vehicles are

Table 2.6. Tours carried out

	1st day	2nd day	3rd day	4th day	5th day
Tour 1	*D-2-D*	*D-2-1-2-2-D*	*D-2-D*	*D-2-D*	*D-2-D*
Tour 2	*D-3-D*	*D-3-D*	*D-3-D*	D-3-3-3-D	D-3-D
Tour 3	*D-4-D*	*D-4-D*	*D-4-D*	*D-4-D*	*D-4-D*
Tour 4	D-8-5-6-D	*D-4-D*	*D-4-D*	D-4-5-9-19-D	D-4-9-D
Tour 5	D-18-13-10-11-13-11-14 -15-15-14-11-14-16-D	D-5-16-14-18-11-10-5 -17-D	D-8-10-14-11-11-11-14-5-D	D-2-18-22-11-11-10-18-14-D	D-10-11-13-16-13-11-11-11-D
Tour 6	D-19-20-22-19-20 -19-20-D	D-19-16-16-11-11-20 -19-D	D-19-20-22-22-22-22- 26-20-19-D	*D-19-22-D*	D-19-28-26-20-19-7-D
Tour 7	D-22-D	D-22-22-25-22-24-22-D	D-22-11-11-D	D-25-25-22-22-24-19-D	D-27-29-27-D
Tour 8	D-25-27-25-D	D-25-25-28-28-26-D	D-25-27-27-24-25-28-28-D	D-27-26-26-28-28-28-20-D	*D-25-22-D*
Tour 9	D-29-D	D-29-D	*D-29-D*	*D-29-D*	D-29-D
Tour 10	D-2-32-26-26-26-26 -32-31-32-30-D	D-28-32-32-26-28-26-9 -28-30-32-5-16-D	D-32-23-13-13-5-9-14-D	D-23-4-5-9-34-32-32-D	D-22-32-D

... symbolizes that a vehicle is in full employment for the 480 minutes. Underlinings indicate that the order was of top priority (loading and unloading of lorries).

Table 2.7. Necessary modifications of routes in comparison with the initial solution

		1st day	2nd day	3rd day	4th day	5th day	Σ
Order completion in accordance with the initial solution	(including cancelled orders)	38	32	21	23	22	136
Allocation of an order to another vehicle	because the originally designated vehicle was at full stretch	–	5	13	11	5	34
	because an order had to be immediately executed (lorries)	2	5	4	3	–	14
	on account of an unacceptable waiting time (>2h) in compliance with the initial solution	3	5	1	2	3	14
	Σ	43	47	39	39	30	198

constantly at full stretch. This means that the plant locations in those sectors will have to be served by other vehicles if additional orders arise. Over and above this it can also frequently be observed that although vehicles are not at full stretch larger orders still have to be handed over to other tours because within the framework of the initial allocation sufficient capacity is no longer available. By comparison, little importance, both in terms of time and quantity, is attached to the loading and unloading of lorries, which according to the description of the problem is of foremost priority. It is, however, precisely these orders which usually have to be dislocated. Deviations from the initial allocation also occur if considerable waiting times (of more than two hours) would arise. Table 2.7 provides a clear classification of the reasons for this sort of changes in routes in comparison with the initial solution.

VI. Evaluation and Interpretation of the Results

In order to evaluate the suggested problem solution, taking the sweep procedure as the basis, it is necessary, on the one hand, to take up travel times required by the fork-lift trucks and the waiting times for the orders into account along with, on the other hand, the actual data collected at the same time during the investigation period. With regard to the necessary travels the corresponding comparison of target and actual times results in a reduction in the amount which has to be travelled of on average 35.7% (see Table 2.8).

This constitutes a reduction in the weekly travel time of about 3 3/4 hours or, at an average speed of 100 m per minute, a reduction of about 23 km. Even

Table 2.8. A comparison of actual and target travel times

	1st day	2nd day	3rd day	4th day	5th day	Σ	∅
Travel times according to the real survey (minutes)	157.4	164.9	113.8	121.2	83.3	640.6	128.1
Travel times according to the suggested problem solution (minutes)	91.5	87.1	75.2	82.6	75.4	411.8	82.4
Absolute changes (minutes)	65.9	77.8	38.6	38.6	7.9	228.8	45.7
Changes as a percentage	41.9	47.2	33.9	34.4	9.5		35.7

with a relatively low percentage of travels in relation to the total utilization time of about 3% a saving of this sort, in view of the reduced costs for wear and tear, repair and for personnel, is of particular interest to the firm.

The effects of the solution on the waiting times for the individual orders after they have been reported to the headquarters are of similar relevance. With regard to the differing times at which orders are reported as well as to the necessity of prompt decisions the waiting times cannot be directly implemented as variables because of the effort connected with a constant plan revision. This, in turn, means that the waiting times cannot be an immediate object of the planning procedure but the obtained result must nevertheless be viewed in the light of this aspect. Waiting times can always be interpreted as delays in the operational progress and generally involve considerable expenses. As far as this vehicle utilization problem is concerned it is to be expected that a reduction in waiting time will result from the diminished travel times to the plant locations. Above all, as a result of the multiplication effect on all subsequent orders in the course of one and the same tour it can be expected that the savings will be on a corresponding scale. Accordingly, the more orders are combined in one tour the greater must be the effect. From Table 2.9 it can be seen that compared with the actual survey, compliance with the derivative tours means that on average every third order can be carried out earlier whereas only every sixth order is subject to delay (often only minimal). The execution times for the rest of the orders remain unchanged. Overall this means both that every order can be dealt with on average almost 12 minutes earlier and that, consequently, the daily waiting times can be reduced by almost 8 hours. The evaluation of these figures must be left to those responsible in the firm, but in terms of a smoothly-running operational progress they are definitely not to be underestimated.

The results obtained during the five day observation and from the evaluation of the data provide altogether noteworthy evidence in favour of replacing the planning procedure which has been practised up to now. In order to stabilize the statements obtained the investigation could be continued over a longer period; simply on account of the scale of the qualitative differences which have already been established between the two procedures it is, however, unlikely that any fundamentally different findings would be made. Apart from that in a more long term application of the sweep procedure a further improvement in the

Table 2.9. A comparison of the real and target execution and waiting times

			1st day	2nd day	3rd day	4th day	5th day	∅
Number of orders executed	earlier	in comparison with the real survey	15	17	11	19	8	14
	later		10	5	10	9	1	7
	unchanged*		18	25	18	11	21	18.6
		Σ	43	47	39	39	30	39.6
Reduction of the waiting times (minutes)		Σ	600	520	270	480	465	467
		∅	14.0	11.1	6.9	12.3	15.5	11.8

* including cancelled orders.

results will be achieved by means of a computer-backed data collection and data control. Up to which problem dimension it is recommendable to computerize the daily utilization control cannot be definitively judged. As far as the problem at hand is concerned this is not necessarily expedient in economic terms. Overall, a rearrangement of the procedure appears to be possible without any significant breaks. The establishment of a demand survey on the day before, e.g. for an extraordinarily high transport demand, also does not constitute any additional work for the management so that the practicality of the procedure appears to be fully ensured.

VII. Outlook

Problem solutions concerning the organization and planning of vehicle utilization when the transport is irregular have hardly been dealt with in the literature. Methods have only been developed for special cases like the determination of the optimum number of vehicles (HEINZ 1967), and in cases where the probability distribution in respect of the transport demand is available (VATERRODT 1975). In this respect the procedure which has been suggested here ventured into an area which has previously hardly been scientifically penetrated, but which in turn makes it all the more relevant for practical purposes. It was not the intention of this study to achieve the very "best" results in the sense of a maximum reduction of costs. The aim was rather to develop evidently improved solutions in comparison with the actual state. In the selection of the planning instruments it was only possible to bring results into play from the more broadly based research in the related area of the standard problem formulation. Simultaneously, this can form a basis of subsequent investigations by means of which it might be possible to further rationalize the internal transport acitivities to be benefit of the overall workings of the firm. Internal company conditions permitting a division of the central depot into various decentralized depots should be considered with a view to further reducing the travel times (see e.g. EBERHARD et al. 1984). A sectoral division in grid squares might be judged to be similarly efficient. Numerous other organization alternatives are conceivable. In every concrete individual case it is, however, necessary to check which organizational design is acceptable. Suitable planning procedures must be determined on the basis of the individual organizational choice.

References

Bodin, L., and Golden, B.: Classification in Vehicle Routing and Scheduling, in: Networks, 1981, pp. 97 – 108.

Bodin, L., Golden, B., Assad, A., and Ball, M.: Routing and Scheduling of Vehicles and Crews – The State of the Art, in: Computers and Operations Research, 1983, pp. 63 – 211.

Carraresi, P., and Gallo, G.: Network Models for Vehicle and Crew Scheduling, in: European Journal of Operations Research, 1984, pp. 139 – 151.

Clarke, G., and Wright, J.W.: Scheduling of Vehicles from a Central Depot to a Number of Delivery Points, in: Operations Research, 1964, pp. 568 – 581.

Domschke, W.: Logistik: Rundreisen und Touren, Volume 2, Munich 1982.

Eberhard, U., Schneeweiß, Ch., and Vaterrodt, H.-J.: Tourenplanung für zwei Depots bei offenen Touren, in: Operations Research-Spektrum, 1984, pp. 39 – 46.

Edmonds, J.: Matroids and the Greedy Algorithm, in: Mathematical Programming, 1971, pp. 127 – 136.

Frieze, A.M.: Worst-Case Analysis of Algorithms for Travelling Salesman Problems, in: Operations Research Verfahren, 1978, pp. 93 – 112.

Gillet, B.E., and Miller, L.R.: A Heuristic Algorithm for the Vehicle-Dispatch Problem, in: Operations Research, 1974, pp. 340 – 349.

Heinz, N.: Untersuchungen zur Ermittlung der kostenoptimalen Anzahl von Transportmitteln beim unregelmäßigen innerbetrieblichen Transport, Ph. D. thesis, Aachen 1967.

Ong, H.L., and Moore, J.B.: Worst-Case Analysis of Two Travelling Salesman Heuristics, in: Operations Research Letters, 1984, pp. 273 – 277.

Paessens, H.: Tourenplanung bei der regionalen Hausmüllentsorgung, Karlsruhe 1981.

Tyagi, M.S.: A Practical Method for Truck Dispatching Problems, in: Journal of the Operations Research Society of Japan, 1968, pp. 75 – 92.

Vaterrodt, H.: Tourenplanung – Modelle und Verfahren, Frankfurt-Zürich 1975.

Chapter 3

Central and Decentral Purchasing in Business Firms

JOACHIM REESE

I. The Problem

In the course of planning the material procurement in industrial enterprises the question recurs regularly whether the organizational prerequisites are at all suitable to support such planning to the necessary extent. The main point of interest here is whether a reorganization of the purchasing activities, by means of either a centralization or a decentralization of these tasks in the various areas, has noteworthy effects on the total costs and earnings situation of the firm.

Generally a whole series of arguments enter into these comparative reflections (see e.g. BAILY and FARMER 1981, p. 16). These arguments range from the economy of the order quantities to the thus influenced holding costs to the altered lead times for the material flow from the supplier to the warehouse as well as to the danger of shortages. Most of these aspects, however, are at the same time also touched on by the ordering policy of the firm, i.e. by the time and quantity of the orders. So it can be presumed that the purchasing organization cannot be judged without taking the determinative ordering policy into account. In turn, it can be concluded from this, contrary to the opinions represented in a number of places (e.g. ORLICKY 1975), that the selection of an ordering policy, in terms of the overall performance of the firm, is not of necessity a largely ineffective and thus unproductive measure. For this reason the main aim of the following expositions, after some fundamental evaluation criteria have been formulated, is to show the way in which the problem of coordinating purchasing activities is connected to the selection of an expedient ordering policy. Secondly, it is examined how conclusions resulting from this influence the choice of the organizational structure and the choice of the order parameters, respectively.

In the face of a multitude of possible comparisons one completely decentralized and one centralized purchasing organization on the one hand and the most well known ordering policies on the other hand are to be singled out and examined here. As a rule, however, this provides enough insight into other means of organization and of planning. As far as the object of replenishment of inventories is concerned a restriction to those materials is permitted which show the important characteristics of a regular demand in a sufficiently high quantity.

II. Criteria for the Assessment of Purchasing Activities

In order to be able to precisely judge the quality of purchasing activities, from an economic point of view, numerous characteristics of the firm's situation and the thus resulting concrete purchasing procedure are to be observed. As a basis for an ensuing evaluation of costs these factors are to be accounted appropriate significance. In as far these characteristics can have properties of a design parameter with regard both to the purchasing organization and to the ordering policy, these are above all:

- the order quantity which on the one hand has to be orientated to the overall demand of the firm and on the other hand to the costs involved;
- the lead time, i.e. the time between scheduling an order and its actual addition to stock, which is a derivative of external market conditions but also of the existing degree of centralization within the firm;
- the purchasing conditions in as far as they depend on the order quantity, e.g. discounts, and over and above that which can be effected by organizational design;
- the safety stock which results from demand uncertainties and which can in addition be reduced by means of a suitable organization of the inventory system;
- the stored quantities which result from the deliveries as well as from the comparative demand;
- shortages which arise as a result of a stock-exceeding demand at a particular time and which cause a reduction in the service rate;
- the personnel demand which, as far as both quality and quantity is concerned, is influenced on the one hand by the organizational design of this area and the functions which have to be carried out within it (see e.g. BAILY 1978, p. 256) as well as on the other hand by the ordering policy and the activity requirements connected with it (see BIGGS et al. 1980).

As one can immediately see, concrete designs of ordering policy and purchasing centralization do not necessarily have effects on these characteristics in the same direction and most certainly not to the same extent. In particular, differing consequences are therefore produced for the choice of an ordering policy according to the existing purchasing organization and for the organizational design of the purchasing activities according to the assumed ordering policy, respectively. In order to be able to quantify these consequences the characteristics are first to be collated and rendered comparable. Expediently the usual cost criteria, which take all the above-mentioned aspects into account, can be formulated for this purpose. Here it is sufficient to limit this to those cost types which are considered in the evaluation of ordering policies, namely replenishing, holding and shortage costs. A restriction of this sort can be made because costs which are solely dependent on the organizational choice are not of interest for a comparative evaluation with regard to various ordering policies.

In the following expositions the task of establishing the specific personnel capacities, which poses considerable quantitative evaluation difficulties because of existing overheads, is to be left aside. Furthermore the premises of a perfect service rate should be given preference which means that shortage costs can be ignored. With regard to the inclusion of shortages in the organizational assessment of a warehousing system EPPEN (1979) can, for example, be referred to. For a single-period model, i.e. without taking a particular ordering policy into account, he reveals cost advantages for a centralized warehousing organization. MEHREZ and STULMAN (1983) modify this result for cases when instead of taking the shortage costs into consideration service constraints are defined; here in the case of a high correlation of the demands of the individual locations a decentralized warehousing system can well be more cost efficient. Finally, considerations concerning overheads can also be left out of account in the following, unlike in multi-factor cases – which have not been dealt with here – in which they have to be considered.

III. The Impact of the Ordering Policy on the Purchasing Organization

As yet mentioned, the object of examinating the purchasing organization is restricted to a comparison of spatially and functionally decentralized purchasing departments of individual production locations with a suitable centralization of these functions. Combinations of these two patterns should be left out of account because it would be mainly questions concerning the internal structure of the branches that would be thus touched upon (see, however, for example BAILY and FARMER 1981, p. 15, or EPPEN and SCHRAGE 1981). Moreover, in order to obtain significant results, it suffices to examine one material which is only required by two locations. A wider distribution of the demand among several production locations will merely serve to substantiate the obtained results.

With regard to the inclusion of ordering policies it is necessary to urgently distinguish between lot-size systems and order-level systems (e.g. NADDOR 1966, p. 47). The notation for the used parameters shall be as follows:

S – order level to which the inventory is filled up after every order;
Q – optimal order quantity (lot size) taking the holding and replenishing costs into account;
s – reorder point upon arrival at which new orders are placed;
t – scheduling period which determines the interval between two ordering decisions.

All manner of parameter combinations are basically conceivable and each determines a particular ordering policy. Attention should, however, be focused above all on the (s, Q)-policy as the lot-size procedure as well as on the (t, S)-policy as the order-level procedure. (The other two elementary ordering policies, namely the (t, Q)- and the (s, S)-policy, are of secondary importance either because they cannot specify every decision variable (NADDOR 1966, p. 12) or because they more or less correspond exactly to another policy (NADDOR 1966, p. 314).) In addition, however, it would seem to be interesting to take several decision situations – with regard to the future course of demand – as a basis for each of these two policies.

The starting point of all further reflections is initially to be the simple, and presumably hardly realistic, question as to whether centralized or decentralized reorders are to be preferred for a case in which the future demand is known in detail and is constant.

Let us at first reflect on the consequences of an (s, Q)-policy, where all the other assumptions of the classic lot-size system should also be valid (see e.g. NADDOR 1966, p. 48). In particular, the abandonment of shortages means that for the reorder point $s = 0$ must be fulfilled. If the easily derivable optimal order quantity Q_0 is considered, the minimal overall costs per unit of time for a centralized purchase clearly amount to

$$C_0^{\text{cen}} = \sqrt{2c_1c_2r} \ ,$$

if r indicates the (constant) rate of demand per unit of time, c_1 the holding costs per unit of quantity and time in stock and c_2 the costs of replenishment independent of quantity per unit of time.

If the overall demand is divided up into locations A and B, i.e.

$$r = r_A + r_B \ ,$$

and if the corresponding replenishing costs were called c_{2A} and c_{2B} (the holding costs remain unchanged, i.e. $c_{1A} = c_{1B} = c_1$, as the inventory system is not to be concerned by centralizing or decentralizing the purchasing measures), then the comparable overall costs for the decentralized purchasing organization amount to

$$C_0^{\text{dec}} = C_0^A + C_0^B = \sqrt{2c_1c_{2A}r_A} + \sqrt{2c_1c_{2B}r_B} \ .$$

For direct comparison of costs the parameters are reformulated as follows:

$$r_A = \varrho_A r \ , \quad r_B = \varrho_B r \ , \quad \varrho_A, \varrho_B > 0 \ , \quad \varrho_A + \varrho_B = 1 \ ,$$

and

$$c_{2A} = \gamma_A c_2 \ , \quad c_{2B} = \gamma_B c_2 \ , \quad \gamma_A, \gamma_B > 0 \ .$$

Now it can easily be seen that centralization must be preferred, if

$$C_0^{\text{cen}} < C_0^{\text{dec}} \ , \quad \text{i.e.}$$

$$\sqrt{\gamma_A \varrho_A} + \sqrt{\gamma_B \varrho_B} > 1 \ .$$

This inequality holds, for example, for $\gamma_A = \gamma_B = 1$ and arbitrary demand proportions ϱ_A and ϱ_B. In this case of constant replenishing costs the positive effect of centralization results solely from the concavity of the optimal cost curve $C_0(r)$. The advantages of a central purchase are further strengthened, when the purchase prices are dependent on the order quantities, e.g. a quantity discount is guaranteed. A possible shorter lead time in a decentralized organization, on the other hand, does not play a role in the establishment of costs as long as the demand is deterministic; the individual orders can be effected later if the lead times are shorter but this does not have an influence on the level of stock on hand because the delivery date remains the same (see NADDOR 1966, p. 31). The safety stock also has no effect on the evaluation of the purchasing organization because if the future demand is known such stock is not required.

Finally, it is necessary to point out that under the same conditions all the ordering policies mentioned at the beginning of this section clearly lead to identical results with the exception of the (t^*, S)-policy when the scheduling period t^* cannot be influenced by the firm. In this case the overall costs are predetermined on account of

$$S = rt^*$$

and result, without the possibility of the planning board being able to intervene further, in

$$C_0^{\text{cen}} = c_1 rt^*/2 + c_2/t^*$$

and

$$C_0^{\text{dec}} = c_1(r_A + r_B)t^*/2 + (c_{2A} + c_{2B})/t^* \ ,$$

respectively. Synergy effects which could yet result from a centralization are thus restricted to perhaps more favourable replenishing costs, i.e.

$$c_2 < c_{2A} + c_{2B} \ ;$$

they cannot be set up by means of the choice of optimum order quantities. But, nevertheless, smaller scheduling periods and therefore more frequent delivery dates would again permit a (discrete) order quantity disposition. Ultimately, $t^* \to 0$ leads back to an (s, S)- or (s, Q)-policy with a correspondingly advantageous assessment of centralization.

Essentially the same results are obtained if a comparative assessment of possible organization patterns is to be carried out assuming that although not constant the future demand remains at least calculable. Though the overall costs are generally higher (see NADDOR 1966, p. 111) the cost relationship between centralized and decentralized purchase does not change. As soon as, however, uncertainty about the future demand prevails new aspects enrich the organizational considerations concerning the purchase of materials. In order to carry this out a stochastic demand course should be assumed for which only a probability distribution is known. The rest of the model premises should be left unchanged.

The derivation of the optimal lot size Q_0 within the framework of an (s, Q)-policy is now based on the fundamental conception that the average inventory on hand results in

$$s + Q/2 - r\bar{t}$$

and the expected number of replenishments per unit of time in

$$r/Q \ ,$$

whereby $\bar{t}$ denotes the lead time and r the expected rate of demand. Moreover, the reorder point can be identified with the maximum demand quantity $y_{\max}$ required during the lead time. Thus, with a central organization total costs arise to the sum of

$$C^{\text{cen}}(Q) = c_1(y_{\max} + Q/2 - r\bar{t}) + c_2 r/Q \ ,$$

the minimum of which with optimal order quantity Q_0 is achieved with

$$C_0^{\text{cen}} = \sqrt{2c_1c_2r} + c_1(y_{\max} - r\bar{t}) \ .$$

The expression $(y_{max} - r\bar{t})$ can obviously be interpreted as a „safety stock", which satisfies possible demand quantities beyond the mean value, and causes corresponding holding costs, the extent of which is considerably influenced by the lead time. If y_{max} can be formulated as the multiple k of the mean demand during the lead time then this results in

$$(k-1)\,c_1\,r\bar{t}$$

for the assessed safety stock. In the case of decentralized purchasing corresponding lead time reductions are yielded on account of the higher degree of flexibility which is to be expected with smaller organizational units (see e.g. BAILY and FARMER 1981, p. 16). At the same time the cost advantage exhibited by the central organization as a result of the more favourable order quantities is reduced by the cost disadvantage

$$(k-1)\,c_1\,(r\bar{t} - r_A\bar{t}_A - r_B\bar{t}_B)\ ,$$

which results first and foremost from the longer lead times but also from the range of the underlying demand distribution. If, for example,

$$\bar{t}_A = \bar{t}_B = \bar{t}/2 \quad \text{and} \quad k = 3\ ,$$

this disadvantage amounts to exactly

$$c_1\,r\bar{t}$$

units of cost. For $k = 1$, i.e. for the deterministic case of demand, it disappears as expected. It also remains below the value which has been derived above if the range between the maximum and the mean demand during the lead time depends upon its length, i.e. when

$$k = k(\bar{t}) \quad \text{with} \quad dk/d\bar{t} < 0$$

is fulfilled. This could be attributed to increased compensation tendencies during a longer lead time. In the special case

$$k(\bar{t}) = 1 + k_0/\bar{t}$$

it is simple to prove the independence of the safety stock from the lead time (NADDOR 1966, p. 150). If over and above that

$$k(\bar{t}) > k(\bar{t}_A) \quad \text{and} \quad k(\bar{t}) > k(\bar{t}_B)$$

is assumed then the presented cost development can change again into a further advantage in favour of centralized purchase.

In practice, a firm frequently employs dynamic heuristic procedures in order to achieve a quick but nevertheless reliable decision. From this point of view the Part-Period Balancing is also very common. The inherent idea, derived from the lot-size rule, says that the holding costs and the replenishing (ordering) costs must be balanced out (for a detailed description see DE MATTEIS 1968 and MENDOZA 1968). Taking the lead times into account, in the case of centralizing all purchasing activities, this procedure causes overall costs

$$C_0^{\text{cen}} = \sqrt{2c_1c_2r + c_1^2(y_{\max} - r\bar{t})^2} + c_1(y_{\max} - r\bar{t}) \ ,$$

which are above the optimum cost of the exact solution in as far as demand uncertainties arise. It can be immediately recognized here that, with regard to a comparison of the organization, different lead times are of even more consequence and that disadvantages in terms of order quantities in decentralized purchasing are compensated for more than is the case with the exact procedure. Finally it is noteworthy, as is later further substantiated with an example (see Table 3.1), that the Part-Period Balancing can even cause a totally different organization recommendation.

If a firm is now committed in its ordering policy to place its material orders in freely determinable but constant periods t, and if it therefore pursues a (t, S)-policy the advantage of a purchase centralization has to be again reassessed. The average inventory on hand is now to be calculated as

$$S - r\bar{t} - rt/2$$

(see e.g. NADDOR 1966, p. 149), whereby the order level S marks the limit up to which additional orders are replaced without necessarily having to use the complete inventory capacity. If shortages do not occur then the necessary level is determined by means of the maximum demand $x_{\max}(t+\bar{t})$ which arises during the ensuing scheduling period t and the previous lead time $\bar{t}$. Thus if a (t, S)-policy is pursued in the case of a central organization the average total costs per unit of time amount to

$$C^{\text{cen}}(t) = c_1[x_{\max}(t+\bar{t}) - r\bar{t} - rt/2] + c_2/t \ .$$

If a demand distribution is assumed for which

$$x_{\max}(t+\bar{t}) = kr(t+\bar{t})$$

prevails this leads to a minimum cost

$$C_0^{\text{cen}} = \sqrt{2(2k-1)c_1c_2r} + (k-1)c_1r\bar{t} \ .$$

For one thing it can be seen immediately that the second term of C_0^{cen} corresponds with the safety stock expression calculated for the (s, Q)-policy so that lead times dependent on the organization in turn cause additional costs. For another thing it is shown by the first term that stochastic demand fluctuations have an even greater impact on the overall costs. This results from the additional weight $(2k-1)$ and is the reason why the order quantity advantage to be expected for central purchasing increases considerably in value. By comparison, the disadvantage of unfavourable lead times is pushed into the background. The plausible explanation for this kind of shifting of weight is to be found in the fact that the uncertainty about the future demand is no longer solely restricted to the lead time $\bar{t}$ but also has to be taken into consideration to the same extent with regard to the scheduling period t.

A look at the (t^*, S)-ordering policy – for which the scheduling period t^* is predetermined, for example, by deliverers – reveals that the advantages of a purchasing decentralization are most likely to be produced under this circumstances. It follows immediately that with

Table 3.1. Exemplary cost determining with varying purchasing data and ordering policies

Purchasing data	Comments	(s, Q)-Policy	Part-Period Balancing	(t, S)-Policy	(t*, S)-Policy
1. $k = 1$	Constant demand No uncertainty; arbitrary lead times	$C_0^{cen} = 200.0$ $C_0^{dec} = 281.5$	$C_0^{cen} = 200.0$ $C_0^{dec} = 281.5$	$C_0^{cen} = 200.0$ $C_0^{dec} = 281.5$	$C_0^{cen} = 250.0$ $C_0^{dec} = 300.0$
2. $k = 1.2$	Uncertain demand fluctuations				
a) $\bar{t} = \bar{t}_A = \bar{t}_B = 1$	Lead times independent of the organization	$C_0^{cen} = 280.0$ $C_0^{dec} = 361.5$	$C_0^{cen} = 295.4$ $C_0^{dec} = 372.7$	$C_0^{cen} = 316.6$ $C_0^{dec} = 413.0$	$C_0^{cen} = 410.0$ $C_0^{dec} = 469.0$
b) $\bar{t} = 1;\ \bar{t}_A = \bar{t}_B = 0$	Increasing lead time advantage with a decentralized organization (applies to b–d)	$C_0^{cen} = 280.0$ $C_0^{dec} = 281.5$	$C_0^{cen} = 295.4$ $C_0^{dec} = 281.5$	$C_0^{cen} = 316.6$ $C_0^{dec} = 333.0$	$C_0^{cen} = 410.0$ $C_0^{dec} = 380.0$
c) $\bar{t} = 1.1;\ \bar{t}_A = \bar{t}_B = 0$		$C_0^{cen} = 288.0$ $C_0^{dec} = 281.5$	$C_0^{cen} = 306.5$ $C_0^{dec} = 281.5$	$C_0^{cen} = 324.6$ $C_0^{dec} = 333.0$	$C_0^{cen} = 418.0$ $C_0^{dec} = 380.0$
d) $\bar{t} = 2;\ \bar{t}_A = \bar{t}_B = 0$		$C_0^{cen} = 360.0$ $C_0^{dec} = 281.5$	$C_0^{cen} = 416.1$ $C_0^{dec} = 281.5$	$C_0^{cen} = 396.6$ $C_0^{dec} = 333.0$	$C_0^{cen} = 490.0$ $C_0^{dec} = 380.0$
3. $k = 2$	Large uncertain demand fluctuations				
a) $\bar{t} = \bar{t}_A = \bar{t}_B = 1$	Lead times independent of the organization	$C_0^{cen} = 600.0$ $C_0^{dec} = 681.5$	$C_0^{cen} = 847.2$ $C_0^{dec} = 1248.5$	$C_0^{cen} = 746.4$ $C_0^{dec} = 887.4$	$C_0^{cen} = 1050.0$ $C_0^{dec} = 1100.0$
b) $\bar{t} = 1;\ \bar{t}_A = \bar{t}_B = 0$	Lead time advantage with a decentralized organization	$C_0^{cen} = 600.0$ $C_0^{dec} = 281.5$	$C_0^{cen} = 847.2$ $C_0^{dec} = 281.5$	$C_0^{cen} = 746.4$ $C_0^{dec} = 487.4$	$C_0^{cen} = 1050.0$ $C_0^{dec} = 700.0$

Underlying data which remain unchanged: $r = 100,\ r_A = 60,\ r_B = 40;\ c_1 = 4,\ c_2 = 50;\ x_{max}(t) = krt,\ y_{max}(\bar{t}) = kr\bar{t};\ t^* = 1.$

$$x_{\max}(t^*) = krt^*$$

and

$$S = x_{\max}(t^* + \bar{t})$$

the optimal costs are

$$C_0^{\text{cen}} = c_1[(k-1)r\bar{t} + (2k-1)rt^*/2] + c_2/t^*$$

and

$$C_0^{\text{dec}} = c_1[(k-1)(r_A\bar{t}_A + r_B\bar{t}_B) + (2k-1)rt^*/2] + (c_{2A} + c_{2B})/t^* \ .$$

Simultaneously, this reveals that there are no order quantity advantages. The lead time as well as the range of the demand make their full impression on the overall costs. The more the lead time can be reduced by means of decentralization efforts the greater the expected cost advantages of this structure. These advantages increase still further with the uncertainty and mean variation of the period demand, respectively, so that ultimately low lead times in connection with considerable demand fluctuations form the best prerequisites for a purchasing decentralization. A reduction of the cost advantages in respect of multiple ordering costs has no effects on the comparative assessment as long as t^* is sufficiently large, i.e. ordering costs per unit of time remain comparatively low.

If the thus obtained findings are now combined into an overall view on the impacts of the ordering policy of a firm on its organizational decisions for the purchasing area then the following consequences must be stated:

The prerequisite for promising considerations about the centralization of the firm's purchasing departments is a demand for the same materials in various production locations of the firm; only then can noteworthy cost advantages in comparison with a decentralized purchase arise which are mainly attributable to more favourable order quantities. If the demand is deterministic these advantages become apparent for all ordering policies (see example 1 in Table 3.1) if one disregards the fact that given scheduling periods and thus implicitly also given order quantities are obviously not able to guarantee these advantages. A serious disadvantage of a centralized purchasing system, namely its comparative rigidity, which is first of all expressed in longer lead times, has no effect as long as demand uncertainties do not exist.

Thus, if it has to be assumed from the start that the demand is stochastic another picture is produced: lead-time advantages let the decentralized organization appear as a more worthwile alternative again. The decision in favour of a particular organization depends now for almost all ordering policies ultimately on the concrete data constellation, whereby reductions in the lead time, even when the demand fluctuations are really small, can have a great effect:

As long as lead times remain independent of the organizational choice centralization will be further advantageous because of lower ordering costs (see the examples 2a and 3a in Table 3.1). Slightly larger lead times become relevant with regard to an organizational comparison in the cases of underlying ordering policies which stress this criterion. Thus, the frequently practised Part-Period Balancing strengthens the trend of decentralized purchasing more than the (s, Q)-policy (see example 2b in Table 3.1). In comparison to the lot-size procedures the (t, S)-policy causes a higher weighting of the order-quantity advantage. This occurs because on account of the observation of the scheduling period a larger quantity is ordered and because the costs arising per time unit during this period increase by a constant multiple. The consequence of this is that the assessment of the central organization improves and with comparable data represents, more than with the other policies, the more suitable alternative (see especially example 2c in Table 3.1). This conclusion is, however, not valid under the premises that the scheduling period can no longer be freely selected but is predetermined by external impacts. In this case of a (t^*, S)-ordering policy special order-quantity advantages do not arise which means that the shorter lead times will possibly become the decisive factor in favour of a decentralized organization (see again the examples 2b and 2c in Table 3.1), in as far as other data, like for example considerable quantity discounts, do not outweigh these advantages.

Finally, as can be also seen from the examples 3a and 3b in Table 3.1, stronger fluctuations of the material requirements intensify all the results summarized above and thereby facilitate the comparative assessment. Thus, the impact of the ordering policy on the organizational choice becomes evidently much weaker.

IV. Outlook

In the course of the preceding section it has been shown that, despite of identical data prerequisites, as a rule with different ordering policies different organization patterns of the purchasing activities rise superior to the rest. The known and practised ordering policies have by no means all been subjected here to an examination of this sort (for surveys of the most important other procedures employed in the practice see, e.g., BERRY 1972, LOWERRE 1975, ORLICKY 1975, or YELLE 1979). Nevertheless, the analysis necessary for this purpose can often be based on the findings presented here about the impact of the individual design parameters. This is quite feasible because all usable ordering policies are more or less modifications of the policies discussed here and as such do not pursue fundamentally different strategies.

A further interesting perspective is produced if the prerequisites which are valid here do not stand up to practical requirements. Above all the perfect knowledge of a demand distribution, possibly even in the form of a continuous density function, appears in the light of the applicability of a method to be threatened by severe criticism. In the presented analyses it is quite possible to do without detailed informations of this sort, which in a dynamic environment is only conditionally reliable and which for this reason can only be used sparingly even if a stochastic demand course is assumed, in as far as only mean and maximum demand figures were explicitly required. This minimum degree of information is, however, not sufficient in every case for a comprehensive cost analysis. It is not sufficient, for example, when backorders are permitted and thus the firm does not necessarily prevent the occurrence of shortages from the cost considerations. Under such circumstances it must be possible to check whether it would be more appropriate in terms of nature of real and human action to dispense with total information in favour of a satisfying cost target (for the general case of satisfying see also SIMON 1979). Presumably it is not necessary to explain that this method would, on the one hand, simplify the organizational choice but that, on the other hand, it can seriously impede and reverse it. Procedures like the Part-Period Balancing throw light on this point of view in that they are not directly orientated to a lowest attainable level of cost but question the attainment of the calculable cost optimum. The possible reversal of the organizational choice has already been seen in Table 3.1.

With regard to an evaluation of the purchasing organization in its dependence upon the ordering policy it would certainly also be interesting to discover how changes in the data related to the ordering policy find expression in the assessment. The results of the preceding section reveal fundamental sensitivity findings. Thus, as a rule, a change in the replenishing costs has a lesser effect on the overall cost function and on the superiority of an existing organization than the proportional correction of maximum demand fluctuations or the lead time. For concrete situations this sort of analysis could or rather would have to be much more precise so that long-term organizational decisions can also have the best possible foundations.

Another aspect, which for similar reasons needs to be examined and which has not been dealt with here, is the real purchasing situation which consists not only of one material and two locations but in both cases comprises a far broader spectrum. This does not, however, necessarily call for several independent studies of the type presented here but rather a more general consideration of numerous dependencies. Thus, for example, if the ordering processes of various materials are linked with each other in form of joint replenishments (for this problem see e.g. Low and Waddington 1967 or Silver 1976) or if despite of independent orders common warehouse capacities are at least used. Only when all these aspects have been processed can an appropriate organizational choice be made which with the recommendation of an organizational design with both centralized and decentralized features can naturally turn out to be much more complex but much more efficient, too.

References

Baily, P.J.H.: Purchasing and Supply Management, 4th edition, London-New York 1978.

Baily, P., and Farmer, D.: Purchasing Principles and Management, 4th edition, London 1981.

Berry, W.L.: Lot Sizing Procedures for Requirements Planning Systems: A Framework for Analysis, in: Production and Inventory Management, 1972, pp. 19–34.

Biggs, J.R., Hahn, C.K., and Pinto, P.A.: Performance of Lot Sizing Rules in an MRP System with Different Operating Conditions, in: Academy of Management Review, 1980, pp. 89–96.

De Matteis, J.J.: The Part-Period Algorithm, in: IBM Systems Journal, 1968, pp. 30–39.

Eppen, G.D.: Effects of Centralization on Expected Costs in a Multi-Location Newsboy Problem, in: Management Science, 1979, pp. 498–501.

Eppen, G., and Schrage, L.: Centralized Ordering Policies in a Multi-Warehouse System with Lead Times and Random Demand, in: Schwarz, L.B. (ed.): Multi-Level Production/Inventory Control Systems: Theory and Practice, Amsterdam 1981, pp. 51–67.

Low, R.A., and Waddington, J.F.: The Determination of the Optimum Joint Replenishment Policy for a Group of Discount-Connected Stock Lines, in: Operational Research Quarterly, 1967, pp. 443–462.

Lowerre, W.J.: Lot Size Rules: A One-Act Play, in: Production and Inventory Management, 1975, pp. 41 – 49.

Mehrez, A., and Stulman, A.: Another Note on Centralizing a Multi-Location One Period Inventory System, in: Omega, 1983, pp. 104 – 105.

Mendoza, A.G.: Mathematical Analysis of the Part-Period Algorithm, in: IBM Systems Journal, 1968, pp. 39 – 46.

Naddor, E.: Inventory Systems, New York 1966.

Orlicky, J.A.: Material Requirements Planning, New York 1975.

Silver, E.A.: A Simple Method of Determining Order Quantities in Joint Replenishments under Deterministic Demand, in: Management Science, 1976, pp. 1351 – 1361.

Simon, H.A.: Rational Decision Making in Business Organizations, in: American Economic Review, 1979, pp. 493 – 513.

Yelle, L.E.: Materials Requirements Lot Sizing: A Multi-Level Approach, in: International Journal of Production Research, 1979, pp. 223 – 232.

Part II

Concepts of Materials Management

Chapter 4

Rational Material Flow Planning with MRP and Kanban

GÜNTER FANDEL and PETER FRANÇOIS

I. Introduction

Shortages of raw materials, the import of products from "low-wage countries", market saturation in many areas, and the burden of costs caused by higher wages and salaries have led to an increase in competition. A company can only survive in this competitive world, if it can successfully reduce production costs.

One possibility of reducing a firm's production costs is by rationalizing its material flow. Material Requirements Planning (MRP) and Kanban are two important approaches towards this objective. Both methods will be described in this essay, subjected to a critical analysis, and finally compared and contrasted.

As criteria for analysing and comparing these methods are taken into account: (1) efficiency in the sense of cost reduction, (2) flexibility with respect to the possible changes in the product mix as well as to unforeseeable disturbances of the production process, and (3) fulfillment of certain requirements of application.

II. Material Requirements Planning

1. The Changeover from the Classical Inventory Models to Material Requirements Planning

Until the 1970s the flow of material was controlled basically by so-called classical inventory policies, provided that optimization processes were employed at all, and order quantities and dates were not left to the judgment of the individual manager – which can still often be found today. In both cases non-existent stocks, or stock levels which are too low, can lead to production downtime, especially where multi-stage manufacturing processes are concerned. On the other hand, the traditional inventory models often lead to increases in stocks, because they are based on previous requirements and assume that past demand rates will occur again in the future so that corresponding amounts of stock are required.

It is, however, quite simple to grasp that parts, which are only required in particular periods, should not be stored during the whole planning period, as is the case with the stock replenishment techniques. It is exactly this type of requirement which occurs case-by-case, often as a result of dependency on superior components, which leads to incorrectness in present stockkeeping models (FOX 1984, p. 44; SWANN 1984, p. 28). This is where Material Requirements Planning can be applied. MRP is a computer based procedure which has the following objectives (TERSINE 1982, p. 287):

- calculation of all input material necessary for the realization of the Master Production Schedule,
- minimization of inventory levels,
- calculation of the periods in which these materials must be ordered (outside purchasing) or released for production (internal production) in order to avoid shortages.

The realization of these aims leads, because of the efficiency of the materials flow, to cost savings, and thus has a bearing on the firm's competitiveness.

It must be noted that MRP is not a completely new procedure which ignores orthodox inventory models, but rather that it is based on those models. A characteristic of MRP is that it takes into account the dependent demand of products which are made up of several components. Another special feature is that the times when this demand arises and when orders must be placed are specified. Time components play therefore an important part in MRP. MRP is a module of computer-based production planning and control systems. Fig. 4.1 shows how it fits into production planning and control systems.

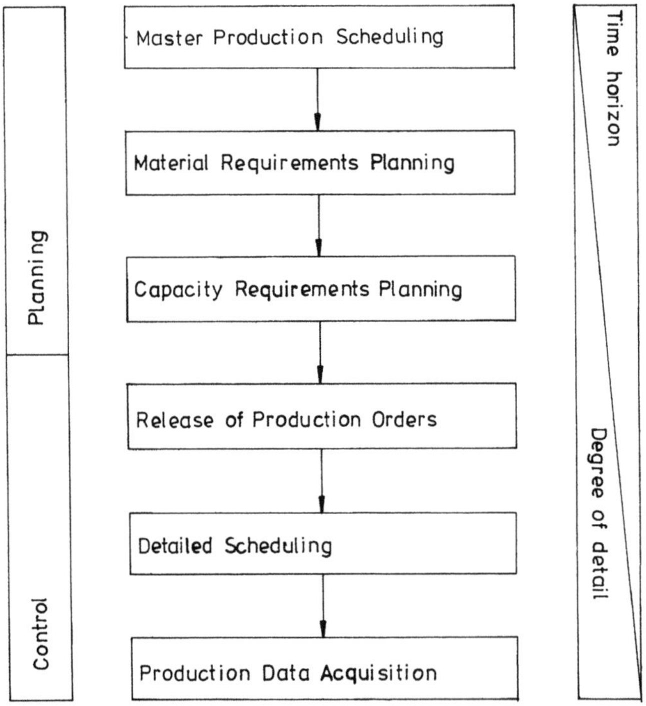

Fig. 4.1. MRP embedded in production planning and control systems

2. Types of Material Requirements

The following distinction can be made regarding the sources of the demand for materials:

- primary requirements,
- secondary requirements,
- tertiary requirements.

Primary requirements are the amounts of final products required and of those assemblies and component parts which are for sale (e. g. spare parts, or the sale of assemblies to other companies). Primary requirements are calculated by adding up customers' orders or by using sales forecasts. Because primary requirements are basically aimed at final products they are only partly material requirements. Secondary requirements are the needs of assemblies and component parts which can be directly determined from primary requirements, and tertiary requirements accrue as indirect requirements in the cost centres. Tertiary requirements, therefore, have no direct bearing on primary requirements. In contrast to primary requirements, secondary and tertiary requirements are used in the company as inputs.

As far as the consideration of available stocks is concerned, a distinction is made between

- gross requirements and
- net requirements.

Gross requirements are the amounts which must be provided in a given period. They include primary, secondary, and tertiary requirements. The needs of gross requirements can be met from disposable stocks, from purchases from outside, or from internal production. Disposable stocks are those stocks which are freely available for use. We will come back to this in more detail later. Net requirements include the supply amounts produced externally and internal production in the corresponding period. They can be arrived at by subtracting disposable stocks in each period from gross requirements.

According to how material requirements are calculated, further subdivisions can be made, namely:

- program-based or deterministic requirements,
- usage-based or stochastic requirements,
- mixed types.

In order to calculate material requirements for each individual period, the program-based calculation uses the Master Production Schedule. Because the structural dependencies between the production plan and material requirements are clearcut, the term "deterministic requirements calculations" is also often used. The determining of program-based requirements for piece-goods is made by means of the bill explosion, and for fluid goods by means of a breakdown of the mixture. The bill explosion will be examined here in greater detail. In the following, all final products, assemblies, and component parts will be described as parts. A final product is made up of component parts or assemblies, and is the result of the last production stage. An assembly is also made up of component parts, or other assemblies, but is itself then incorporated into assemblies or final products. Component parts are materials bought in from outside and which have not yet gone through any production stages in the company.

Usage-based or stochastic requirements calculating starts with the amounts used in the past, and then uses forecasting procedures to determine material requirements. The method is described as stochastic because material requirements are not calculated exactly but are dependent on chance, in spite of the fact that the production plan is known.

A further possibility is a combination of program-based and usage-based requirements calculations. It is employed, for example, when some requirements are derived from primary requirements (secondary requirements), while at the same time other requirements are tertiary requirements.

3. Calculating the Amounts of Material Requirements and the Times for Ordering Materials

(a) Usage-Based Versus Program-Based Material Requirements Calculating

Usage-based calculating of the gross requirements is less difficult than program-based calculating, but it is also less exact. This is especially so when the primary requirements alter with the lapse of time. Actual material requirements then differ considerably from previous values. In order to avoid shortages as far as possible this inaccuracy must be countered, where usage-based material requirements are concerned with increased safety stocks, which in their turn lead to increased stockkeeping costs.

Instruments for usage-based requirements planning are the time series analysis and forecast calculations. With the help of the time series analysis it is examined as to which type of data alterations exist, given the corresponding use of material in the previous periods. In the forecast calculation a forecast procedure is employed which is suitable for the calculated data alteration.

Compared with manual planning EDP offers the possibility of more program-based, and therefore more exact planning. Because industrial companies often have many assemblies and components to manage it is necessary to decide for which type of material, with requirements directly dependent on the production program, it would be worth accepting the increase in calculations which is necessary for using the program-based calculation, and for which type of material the usage-based calculation is to be employed. This decision can be made by using the ABC analysis (DICKIE 1951, pp. 92–94).

Because inexact requirements planning for classes A and B leads to high stockkeeping costs, depending on their high value, requirements for these classes must be planned as exactly as possible. For this reason, program-based requirements calculating is used for these types of material. It is recommended that the less complicated usage-based requirements calculating is used for materials of class C. Of course, suitable safety margins must be considered, because even materials with low usage value cause high costs when shortages occur (e.g. produc-

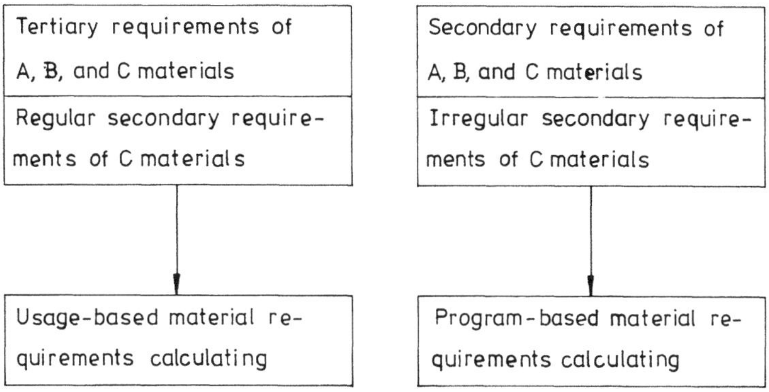

Fig. 4.2. Relationships between the types of material requirements and calculating methods

tion breakdown). Apart from this, those material types of class C which are needed fairly irregularly, or even only once, should not be planned past-oriented, because the corresponding stocks will then be kept too long, and will lead to unnecessary capital costs. The relationships between the different types of material requirements and the type of calculation belonging to them is shown in Fig. 4.2.

Now that the areas in which usage-based and program-based requirements calculating, and their advantages and disadvantages, have been shown, it is proposed to discuss program-based material requirements calculating in more detail.

(b) Program-Based Requirements Calculating

Program-based requirements calculating needs as input the primary requirements, information as to bills of materials, information on the expected amount of rejects or shrinkage, setup and lead times, information on available stocks, and on manufacturing orders which have already been made or on commissions which have already been cleared. Apart from this, before MRP can be carried out it has to be decided which materials are to be produced internally and which are to be purchased from outside.

In order to avoid the repeated carrying-out of MRP for parts which are found at several stages of production, or which form part of various superior parts, the procedure is based on scheduling stages – starting with scheduling stage 0 – and not on stages of production. Numeration starts with the final product and works its way backwards to the material. Although a part can be found at various stages of production this is not the case as far as scheduling stages are concerned, because the scheduling stage for a part is the lowest stage of production at which that part is encountered during production. By proceeding on the lines of scheduling stages, the amount of calculating to be done is also reduced, because less data have to be processed.

(ba) Calculating Gross Requirements

The first step in program-based material requirements planning is to calculate the gross requirements, i.e. how many of a particular part must be supplied in a period. First of all the primary requirements are taken from the Master Production Schedule.

Gross requirements of assemblies and component parts follow from the following calculation:

  primary requirements
+ secondary requirements
+ usage-based requirements
+ margin for inaccuracy

= gross requirements.

Primary requirements for assemblies and component parts only exist when these are to be sold (e.g. spare parts). Secondary requirements are derived from the requirements of final products and of superior assemblies in the course of the program-based requirements planning. Usage-based requirements for component parts and assemblies can occur when these belong to the C parts group, or are used as cost centre material (tertiary requirements). This type of requirement is transferred from the usage-based requirements planning to the program-based requirements planning so that, when the secondary requirements for inferior parts are calculated, all requirements for superior parts can be considered using the bills of materials. If program-based and usage-based requirements occur parallel to one another, care must be taken to avoid duplicating the calculations. This can be prevented by using only those amounts of material as a database which were disposed on a usage basis in the previous period. Gross requirements are similar to all other amounts in material requirements planning in that they are differentiated from period to period.

(bb) Calculating Net Requirements

Once the gross requirements have been calculated the net requirements are found as follows:

  gross requirements
– stock on hand
+ safety stock
+ margin for inventory shrinkage
– released production of internally-produced parts
– outstanding orders for bought-in parts
+ reservations for released commissions of superior parts

= net requirements without rejects

+ margin for rejects

= net requirements

Those parts which are actually in store are the stock on hand. Inventory information is stored in the parts batch. To compensate for unforeseen changes of requirements, minimum inventory levels are built in; the amounts of these levels must be calculated for each part. Any possible stock shrinkage can be taken into consideration by means of a margin. Net requirements also take into account the fact that at certain times units of quantity go into stores which are the result of purchase or manufacturing orders, which have already been released. Expected rejection percentages must be subtracted from those units of quantity, which come from net requirements calculated before the start of planning, because rejected parts cannot be brought in to supply gross requirements. The lead times of the commissions released before the start of planning and the delivery times of the indents put in before planning started lead to these materials arriving in later part-periods; MRP takes this into account by means of the time components.

Manufacturing commissions which have already been released mean that each inferior part is reserved, and this increases net requirements. An increase in net requirements can also occur by reason of rejects in production, or through bought-in parts. Reject parts are those parts which cannot be used in the next stage of production because of defects in quality. The rejection percentage is derived from empirical values, and is also stored in the parts batch. The difference between gross requirements and net requirements without rejects is known as freely available or disposable stock. In the literature the reservations for released commissions for superior parts are often not taken into account in the definition of freely available or disposable stock, in contrast to the definition used here. If, however, the released manufacturing commissions are to be counted as part of available stocks, the reservations for released commissions for superior parts must be considered when available stocks are being calculated. Reserved parts cannot be used, as this can otherwise lead to production hold-ups and those commissions which have been cleared – which count as disposable stocks – will not then go into stores at the time planned.

If net requirements have been calculated it is known which quantities are actually needed at which points in time. If the freely available stock exceeds gross requirements then net requirements are zero. The difference between freely available stock and gross requirements is carried over in this case to the next period as stock on hand.

(bc) Determining Lot Sizes

Where positive amounts of net requirements exist, net requirements planning is followed by measures towards the supply of needs. Where bought-in parts are concerned, orders have to be made, and when it is a case of internally produced parts, manufacturing orders have to be put in. In order to minimize the amounts of setup costs or fixed order costs and stockkeeping costs, net requirements in both cases are pooled into lots. MRP software usually offers the options shown in Table 4.1 for the calculation of the lot sizes.

In so-called lot-by-lot ordering the lot size is exactly the same as the net requirements. This method only takes the minimizing of stocks into account. Materials are supplied exactly at the times they are needed. If fixed lot sizes are

Table 4.1. Lot size procedures in MRP software

Lot size calculating without optimizing	Optimizing procedure
Lot-by-lot ordering	Economic Order Quantity
Fixed given lot sizes	Least-unit-cost algorithm
Combined amounts required in several periods	Part-period algorithm
	WAGNER-WHITIN algorithm

given, or if the amounts required in several periods are combined, no exact lot minimizing takes place.

The standard software usually offers the Harris formula and the least-unit-cost algorithm as optimizing processes.

The part-period algorithm is implemented less often. Dynamic programming (WAGNER and WHITIN 1958, pp. 89–96) is scarcely employed, although it might well be the most suitable procedure. The use of the approximation procedure, and especially the frequent use of the Harris formula, is explained in practice by pointing to its simple applicability and to its low cost-sensitivity in the optimal area (MÜLLER-MERBACH 1962, p. 79; EILON 1962, p. 244; NADDOR 1966, p. 52; DINKELBACH 1969, p. 32). On the other hand, the fact must be taken into account that simple applicability hardly matters in the light of the use of EDP with its much shorter calculating times. For this reason dynamic programming should be integrated more often into software packages.

Further disadvantages of lot size planning within MRP are the following points:

- The multi-stage production is not taken into account. This is because up to the present no satisfactory approaches to multi-stage lot size planning exist (SCHEER 1978, p. 199). In single-stage lot size planning the lots of superior parts are formed solely in consideration of their setup and stockkeeping costs. Lot size planning for superior parts means, however, that the inferior parts are needed earlier, and that their amounts needed can change in the individual periods.
- Through continuous lot size planning throughout the scheduling stages the future requirements for inferior parts are drawn more and more into the present, so that pressure on capacity is increased. This effect is strengthened with the number of scheduling levels, and can lead to existing capacity being insufficient for the production of these lots. Exceeding capacity can be dealt with, up to a point, in the following planning stages only by adjustment measures which, however, lead to higher costs. This problem could only be overcome by simultaneous lot size planning at all scheduling stages.
- Lot sizes for parts used by the same production facilities are determined in isolation. The fact that the lot sizes are in competition for machine scheduling times is not taken into account. Becauce production cycles for the lots are not coordinated on a time basis, this can lead to batch frequencies being determined which cannot be realized, although there would, on the whole, be

enough capacity (the problem of sequencing in lot size planning). There is no guarantee that enough production facilities are free for production of a part as soon as the stocks of this part are nearly exhausted.

- Because of the isolated lot size calculation sequence-dependent setup costs cannot be taken into consideration (scheduling problem in lot size planning). In contrast to the sequencing problem, the scheduling problem leads to suboptimal setup times and setup costs, whereas the sequencing problem leads to unrealistic machine occupancy plans or production cycles. The sequencing problem can also occur when the setup costs are independent of the sequences. Setup costs dependent on the sequences accrue when the extent of the fixed batch costs depends on which type of part is to be exchanged for which other type of part. This is of importance, for example, in dyeing procedures in the textile and paper industries, or in rolling mills in the steel industry.
- A further problem of lot size formation is that the setup costs are not known exactly when there is a choice as regards machine occupancy. This is because the decision regarding machine occupancy is made after MRP has been carried out.

Along with the theoretically unsatisfactory lot size optimizing in MRP the missing decision criteria as regards internal production and outside purchasing can also be seen as disadvantages of computer-based material requirements planning. A decision regarding internal production and outside purchasing must be made in advance. The fact that this decision is often dependent upon capacity use in the company cannot be taken into consideration, because capacity use is not known at the time material requirements planning is carried out, and is only calculated later in the Capacity Requirements Planning.

With the formation of the order lots the computer-based material requirements planning has already taken over a part of the functions of the procurement department. Information about lot sizes and the requirement times of the parts which are to be procured is passed on to the purchasing department. The corresponding orders are then made after choosing the suppliers and taking delivery times into account.

(bd) Forerunning Time Displacement and Bill Explosion

Information on the amount of materials and the times when the parts which are to be produced are needed is relevant to production scheduling and control. With multisectional products, requirements and times for inferior parts can be deduced from the calculated lot sizes. The types and the numbers of inferior parts are then stored in the parts' records and the parts' structure records. The parts' records also contain the lead times of the parts, and the parts' structure records contain the forerunning times. A forerunning time of t units for the superior part means that the inferior part will only be needed after the superior part has already spent t units of time in production. The forerunning time displacement can be calculated from the lead times and the forerunning times as follows:

  lead time of the superior part
– forerunning time of the superior part as opposed to inferior part

= forerunning time displacement

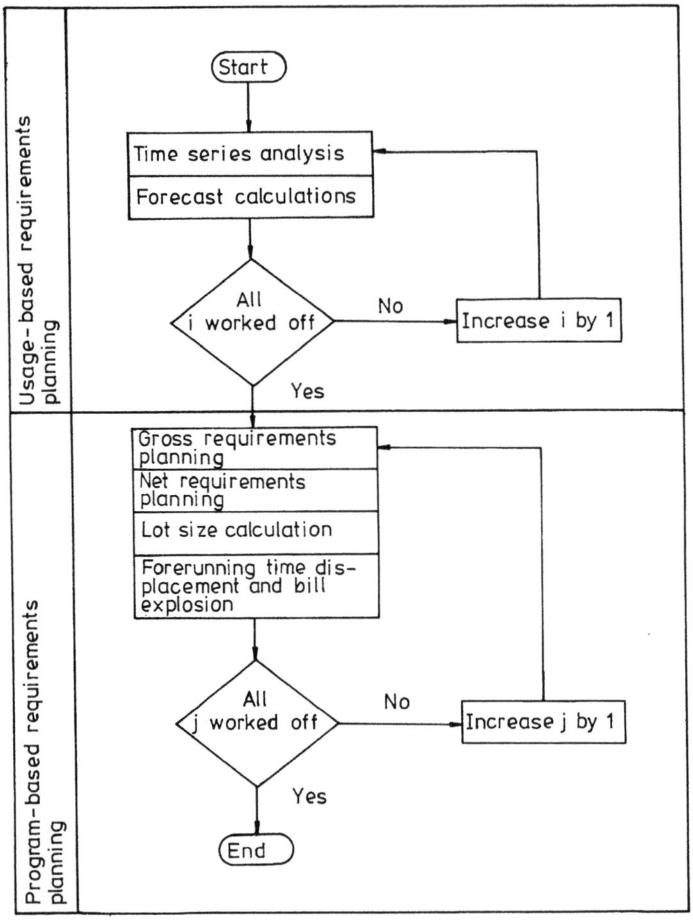

Fig. 4.3. Planning stages for computer-based material requirements

Key: i is the index of those parts with a usage-based schedule;
j is the index of those parts with a program-based schedule.

A forerunning time displacement of t units of time means that the inferior part will be required t time units before completion of the superior part. Consideration of the forerunning time displacement in MRP serves towards the consideration of dependencies to time along with quantitative dependencies, in order to avoid production stoppages caused by ignoring lead times and time displacements.

If the lots of the superior parts are multiplied by the production coefficients, and the lead time displacement is taken into account, the secondary requirements of inferior parts are obtained. These secondary requirements are in their turn used to obtain the gross requirements of inferior parts. This cycle, which can be seen in Fig. 4.3, is run through all scheduling stages until all those parts have been gone through for which program-based scheduling exists.

4. The Development of MRP to MRP II

Up to now materials requirements planning has been a production planning and stock control system which is able to work out

1) which materials,
2) in which amounts,
3) are required at which times and
4) when they must be ordered or manufactured.

The plans can be revised, if unforeseen changes occur in the data. The basic premise is, however, that there is always enough capacity available. The plan produces external and internal orders without taking the possibilities of realizing them into account.

If the assumption of available capacity is not given – as is often the case in industrial concerns – the production program can either only be realized at an increase in costs, or even is impossible to carry out. This is the reason why it must always be asked whether enough capacity, manpower or production facilities are available (SWANN 1984, p. 30). At first, in the Master Production Scheduling a rough estimate is made of the required capacity. After MRP has been carried out, Capacity Requirements Planning takes all the parts which have to be produced into account. If discrepancies occur between required and existing capacity, measures can be taken to increase or to reduce capacity, or, on the other hand, primary requirements can be revised. In this way a closed feedback system is obtained in which MRP is an important module of the overall system. This system is also known as Manufacturing Resource Planning, in short, MRP II. Definitions of MRP or MRP II are not always uniform, neither in theory nor in practice. On the basis of what has already been said here it can be stated that MRP should be employed, in the main, when the final product is expensive and has a complex structure, i.e. is made up of several parts. The requirements for inferior parts must be dependent in a foreseeable manner on the requirements for superior parts (TERSINE 1982, p. 325).

A study published in 1981 on the application of MRP in American companies showed the following average amount of success for MRP (ANDERSON et al. 1981, p. 46):

- the inventory turnover increased by 50% following the introduction of MRP,
- delivery lead times decreased by 17%,
- punctuality of deliveries rose by 55%.

The efficient working of the system and its profitability depend to a great extent on how up to date the data are, and how complete they are, and on their accuracy. These requirements for MRP must be taken into acount. For example, inventory informations must always be kept at the latest level, dependent on constant addition to stock and withdrawal from stock, so that the calculations of the amount of material required can achieve reasonable results. Considerable improvements can be made to the altering service for stores data by employing interactive processing, databank technology and production data acquisition. It is exactly because MRP has to cope with enormous amounts of data and is the starting point for further planning and control modules, that poor quality input leads quickly to the failure of the whole production scheduling and control system.

Now that material flow control on the lines of MRP has been illustrated it is proposed to follow this with a detailed analysis of material flow control using the Kanban system.

III. The Kanban System

1. Introduction to the Kanban System

The Kanban system was developed at the beginning of the 1950s under the aegis of the then Vice-President of Toyota, TAIICHI OHNO (MONDEN 1981 a, p. 46; SUGIMORI et al. 1978, p. 185). Basically, Kanban has the following objectives:

- a reduction of material stocks between the production centres and in stores,
- a reduction in the manufacturing cycle times,
- an increase in flexibility regarding short-term readiness to deliver,
- an increase in labour productivity.

The starting point for OHNO's considerations was the organization of material supplies on a supermarket basis. This means that the consumer, or the consuming centre, takes a part from the shelf when it is needed, and the shelf is always restocked when a gap occurs, or when stocks are lower than a minimum level.

2. How the Kanban System Works

The instrument of the Kanban system is the "Kanban" (= card, label). Material flow control within the Kanban system takes place using this data carrier. Basically, two types of Kanbans are used (MONDEN 1981 b, p. 29):

1. the withdrawal Kanban,

and 2. the production-ordering Kanban.

Withdrawal Kanbans are used between the consuming areas and the buffer stores, and the production-ordering Kanbans are used between the buffer stores and the production centres.

Production control using Kanbans is one possibility – but not the only one – of how to use information techniques to bring about just-in-time production. Another possibility might be, for example, using EDP. Just-in-time cannot, therefore, be equated with Kanban.

Fig. 4.4 shows how the withdrawal and production-ordering Kanbans are used in order to control just-in-time production.

It can be seen that every production centre is at one and the same time a producing point (material source) and a consuming point (material sink). The following steps explain how the feedback control system works.

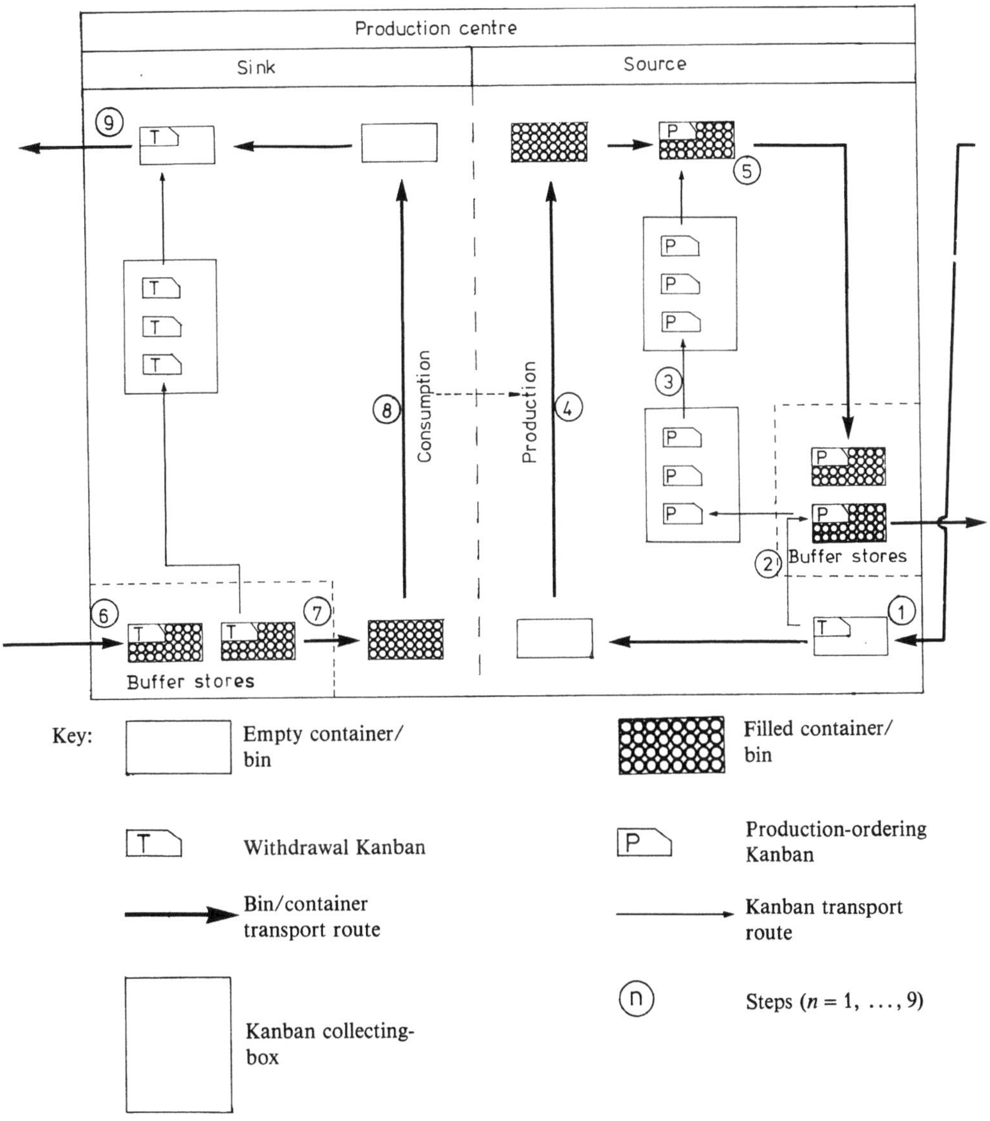

Fig. 4.4. An example for a Kanban system

Step 1:

Starting point of the Kanban system is always the consuming section of the production centre. From here a worker goes to the previous production centre with the necessary amount of withdrawal Kanbans and the empty bins. He leaves the empty bins there (in Fig. 4.4 one bin).

Step 2:

The worker then adds withdrawal Kanbans to the filled bins which he takes from the buffer stores. He removes the production-ordering Kanbans, which he deposits in the Kanban collecting-box. When the withdrawal and production-ordering Kanbans are exchanged they must be carefully cross-checked, otherwise breakdowns in material supply could occur later.

Step 3:

After short intervals, e.g. every hour, the Kanban collecting-box is emptied. The production-ordering Kanbans are the manufacturing orders for the centre, and are deposited in the order box. They are dealt with in the order in which they arrived.

Step 4:

The material which was requested using the production-ordering Kanbans is produced and placed in empty bins.

Step 5:

The re-stocked bins are furnished with the relevant production-ordering Kanbans and brought to the buffer stores so that the material is available and can be withdrawn by the subsequent production centre.

Step 6:

In order to be able to produce the material which is brought to the buffer stores, the production centre itself needs material from the previous centre. These input materials are stored in a buffer store.

Step 7:

Before the materials are used, the withdrawal Kanbans are taken from the bins and placed in a collecting-box.

Step 8:

During the production process the material is removed from the bins.

Step 9:

When a sufficient number of withdrawal Kanbans has been accumulated in the Kanban collecting box, a worker takes the withdrawal Kanbans and the bins to the previous production point, gets a supply of full bins, and triggers off a demand cycle as seen already in step 1.

The steps described here can be found in or between all production centres of a Kanban system, whereby the last production centre receives its orders from the production planning department, and the first centre can pass on the orders to suppliers by using Kanbans. Kanban control between supplier and buyer works on the push-principle; this means that the necessary parts are delivered at fixed times. The internal Kanban control loops work on the pull-principle, except as regards the finished goods stores. The above-mentioned buffer stores are necessary because the pull-principle means that successive production centres each have to get their supplies from the previous centre, and because smaller production hitches have to be balanced out. Buffer stores are kept to a minimum, because Kanban works with small lot sizes (RODAT and SPÖRL 1984, p. 323; SOOM 1986b, p. 448). The pull-principle used in Kanban is often described as a drawing or towing principle, because a production order is given to the last production centre, which itself, by placing orders step by step with each of the preceeding centres, draws further production centres behind it.

3. Preconditions for the Use of a Kanban System

The Kanban system can only be used when the following requirements regarding the organization and the workforce either already exist or can be met:

(a) Constantly Repeated Production Sequences

It is necessary that each phase of the operation, or the parts which are to be produced, are constantly repeated, in order to avoid having to renew the Kanban schedules constantly, and to be able to use the Kanban cards again and again. This is also known as repetitive production sequences. These are especially to be found where many standardized parts are used and the number of product variants is low. These requirements should be taken into account at the construction department, in particular by means aimed at standardization and uniformity. It can be seen, for example, that there is a trend in the Japanese car industry towards equipment packages or to so-called complete packages, in order to keep the numbers of product variants down. In order to have constantly repeated production sequences it must also be guaranteed that the production program is not subject to any radical deviations as regards the type and the amount of products manufactured (MONDEN 1981a, p. 40). Japanese car manufacturers fulfil this type of precondition by manufacturing for a period of 10 – 30 working days to a constant production program with the same model mix each day (RODAT and SPÖRL 1984, p. 323). The Kanban system would break down without the precondition of a smooth, harmonized demand for the products. Even minor increases in the amounts to be produced can mean that the same daily sequence can only be brought to an end by means of overtime, because an increase in the rate of production is achieved through an increase in the batch frequency and not through an increase in the lot size (MONDEN 1981a, p. 43).

In order to determine which parts are suitable for the application of Kanban the ABC analysis can be used. It would pay to use Kanban control for parts with high or average usage values – A or B parts – because keeping these parts in stores leads to a much higher cost of capital than with C parts. The ABC analysis can be extended by using the XYZ analysis, which subdivides the material into those with regular, constant usage (X materials), fluctuating usage (Y materials), and irregular usage (Z materials). Kanban should then be used for those parts which are not only A or B parts, but also are used on a constant or fluctuating basis. All those parts which belong to group C or are only required irregularly should not be controlled using Kanban. Parts which because of their size are not suitable to be stocked in buffer stores are also unsuitable for Kanban control. It can be seen that it can certainly be efficient to use Kanban only for a particular group of parts, or only in certain production areas.

(b) Layout of Production Facilities in Line with the Production Flow

In order to avoid long distances between the production centres the machines must be installed where possible in line with the flow of production (MONDEN 1981 a, p. 39), whereby assembly islands can exist in certain production areas. Apart from this, production facilities must be coordinated with one another as regards capacity, and the transport routes and spaces for the buffer stores must also be taken into account. The harmonized production process determines the coordinated rhythms of all production centres and islands. However, setting up the production facilities in accordance with the production flow is not an imperative assumption for the success of Kanban. For example, Kanban control loops might also be realized in a job shop manufacturing. This would, however, lead to larger transport routes and would be less easy to keep under control.

(c) Production Processes, in Which Kanban Can Be Used

Production processes can be divided into continuous and discontinuous processes. Kanban is not suitable for continuous processes, e.g. wire or paper production, because here an unbroken series of production units is generated. In contrast, the discontinuous process means that the production sequence is interrupted after one or more units have been produced, or at least can be interrupted. Kanban would probably seldom be suitable for a discontinuous process with en bloc production (batch production), because in most cases the batch sizes often imply large lots, for technical and economic reasons (e.g. kilns in earthenware manufacture). If unlimited production exists, i.e. the production sequence can be interrupted after the manufacture of each single unit, this is an essential criterion for the use of Kanban.

If production is classified according to the way in which orders are placed, distinctions can be made between single-order production, serial-order production and supply production. Single-order production seldom has repetitive production sequences, and when they are found then only on a small scale. Kanban can, therefore, only be applied to these repeated parts, and then only if the parts are not required on a completely irregular basis. Serial-order production fulfills the precondition of repetitive sequences, but mostly for the term of the order only. Here it must be decided whether it is efficient building up a Kanban control system for this type of order size. This will certainly not be the case where the production of smaller lot sizes is concerned. Kanban is best suitable for supply production, and here especially for large-scale and mass production because here the number of repeated activities and sequences is at its highest.

(d) Workers' Training and Motivation

A definite precondition, for production and technical reasons, is that the employees carrying out production must be able, and prepared, to master more working processes, and run more machines, than is usually the case with assembly line production. If the workers have mastered several processes they can help each other in case of illness, and otherwise stand in for absent colleagues. This is necessary in order to guarantee, or keep up, the constant flow of material necessary in the Kanban system because of the low amounts of stocks between the production centres. The employee has to have more independence, more responsibility, and a deep sense of team spirit. The workers in a production centre have to control their own stock levels, and are also responsible for the supplies of material.

This increase in the amount of responsibility delegated to the workforce leads to an increase in motivation which can also be influenced by improved wage systems and bonus payments. An increase in productivity can be achieved by an increase in workers' motivation. The competence to handle the new system, and the ability to operate several machines and several working processes means extensive and intensive training programs for the workforces. The question remains unanswered for the present as to how far the relationship of Japanese workers to their companies and their employers, so necessary for the success of Kanban, exists in other countries or can be transferred to other countries. Along with cultural and sociological factors, loyalty to the company is strengthened by the custom of guaranteeing, in general, a job for life, and is of great importance to the proper functioning of Kanban.

(e) Quality Improvement and Quality Control

A further precondition for the successful employment of Kanban control is the high quality of the parts delivered to the following production centres. If a production stage has manufactured parts which, because of their poor quality, cannot be sent to the next stage, or cannot be used there as inputs, production standstills are caused if the shortages cannot be met from the buffer stores between the production centres. Frequently, the standstills not only affect the next centre but also have a wave effect on other centres.

The demand for an improvement in quality goes hand in hand with an increased amount of quality control, which can be realized either by the production workers themselves, using additional control processes, or by automatically controlling the production process (WILDEMANN, p. 53). Automatic quality control can take place, for example, by using scales, microswitches or sensors which are built into the production facilities and which can compare the realized values with the target values, after taking tolerances into account. If the quality is no longer satisfactory, this is reported optically, or by means of a acoustic signal, or the machine is switched off automatically. A high level of investment is often necessary for automatic quality control, so that an analysis must be made in every case, whether the automatic quality control or quality control using the workers saves more costs.

(f) Smaller Lot Sizes by Reducing Setup Times

With the realization of Kanban an attempt is made to achieve by reducing the lot sizes lower inventory levels, shorter lead times for the orders, and an increased flexibility as regards the immediate readiness to deliver. Achieving these objectives only makes sense, however, when the improvements aimed for are not overcompensated by additional setup costs for the smaller-sized lots. Apart from this, it must be taken into account that although smaller-sized lots can reduce lead times of individual orders, on the other side they do, in fact, lead to longer lead times, given the same conditions because the amount of setup time increases in relation to the processing time. When Kanban is introduced, therefore, production setup times must be worked on increasingly. If shorter setup times can be brought about, this increases the company's flexibility, because the change-over from one product to another, or from one variant to another, can take place very quickly. Shorter setup times mean smaller lot sizes, which in their turn lead to lower volumes of buffer stocks (SOOM 1986a, p. 363).

A reduction in setup times can be achieved, for example, by attempting to carry out as many setup processes as possible while the machines are running, or to use specially trained crews to do the changing-over as quickly as possible. Further possibilities are the purchase of new machines which can carry out the change-over themselves – e.g. changing tools – or which can be changed-over very quickly, or to rebuild the existing machines. As far as reducing of setup times and costs are concerned it must be noted that this is an important task of industrial management, even without the introduction of Kanban. For the introduction of Kanban, however, additional efforts are necessary in this area, because Kanban cannot be employed economically in processes with long setup times (SOOM 1986b, p. 449).

The list of the preconditions for the employment of Kanban shows that the introduction of the feedback and control system requires investment and learning processes. The question is, therefore, in which areas this system can be used economically. Up to the present experience with Kanban in Germany has mainly been made in the electrical goods industry, in car making, and in the mechanical engineering industry. Kanban control is, however, also used by companies in the cosmetics industry, by manufacturers of furniture, household goods, lamps, glasses, and by companies delivering to these branches (WILDEMANN 1987, pp. 55).

4. Costs and Benefits of Kanban Control

When the Kanban system is introduced costs for the following measures will accrue:

- calculating those areas of production which, on the basis of the above preconditions, are suited to the introduction of Kanban control,
- altering the layout of the production facilities, whereby the spaces for the buffer stores and the transport routes must also be planned,
- training for the production management as regards the functions and planning of the Kanban system,
- organizational alterations to production scheduling as regards the construction of the feedback control systems, the harmonizing of capacity, choice and design of cards etc.,

- training for the production workers as regards the function of the Kanban system and the operation of additional working processes,
- changes in the wage system,
- additional efforts and investment in the area of guaranteeing quality and reducing setup times.

Along with the costs of bringing the Kanban system into operation, the following disadvantages have to be accepted:

- The Kanban system is not suited to major changes in the rate of production.
- It is hardly possible to make items to customers' specific orders with relatively short delivery times, because a production plan must exist with the same daily sequences over a long period time.
- Technical or other hitches in a production centre or in a production facility spread very quickly over the whole production area, because of the low volumes of buffer stocks, so that delays or shortfalls in production occur (RODAT and SPÖRL 1984, p. 323).

On the other hand there are many beneficial aspects to the use of the Kanban system:

- Stocks of materials are low while products are usually finished at due time and readiness to deliver is high. In this way the costs of investment in stockkeeping and in stores administration are reduced.
- Lead times for orders are shorter.
- Less control is necessary as the system is basically selfregulating. This means that expenses for data processing and especially for Production Data Acquisition are lower.
- The workers are more motivated, which leads to an increase in productivity.

In practice floating assets should be reduced by about 50 to 70 percent and lead times by about 30 to 50 percent. The increase in productivity should amount to more than 25 percent (WILDEMANN 1987, pp. 54 and 58. Figures are based on 100 pilot projects in which the Kanban system was used).

IV. MRP and Kanban – Differences and Joint Characteristics

Both material flow control systems, MRP and Kanban, have now been illustrated and subjected to a critical analysis. In this chapter the systems will be compared and contrasted, and their differences and joint characteristics will be listed:

- One of the differences between the two systems results from their different origins. The predecessors of MRP are the so-called classical lot size systems and stock replenishment policies. Kanban control was developed from the supermarket principle, by taking this idea from the area of trade and carrying it over to material flow control in production. Buffer stores replenishing in Kanban can be compared with the order point system, which belongs to the stock replenishment policies.

- Both Kanban and MRP can be used for the procurement respectively supply of internally produced parts and those purchased from outside.
- Whereas MRP calculates the amounts of materials and the corresponding dates with a view to the future, the Kanban system determines the amounts and the dates only after a certain amount of material has been consumed. On a very short-term basis, therefore, material requirements are met through production or procurement which has already taken place. Material supply in Kanban is therefore not made on a just-in-time basis, which might be thought to be the objective of Kanban. After production procurement, the parts spend a certain amount of time as buffer stocks and then are processed further in the next production stage. In contrast to this, MRP seeks to avoid buffer stocks and to order the materials at the exact point in time when they are required at the next production stage. In principle, MRP aims at just-in-time production or just-in-time supply. Because it is based on materials already used, Kanban has to put up with the disadvantage of buffer stores to meet the needs of the next production stages. In addition, inventory safety factors have to be considered for the buffer stores. One advantage is that the production centre knows the exact amount required with Kanban, because this is the amount of parts used. With MRP the amount of actual future demand cannot be determined exactly, because forecast errors and target-performance deviations as regards rejects, shrinkage etc. can occur. For this reason margins are used to avoid production standstills as much as possible, and this leads likewise to the build-up of stocks.
- When MRP is used the supply of parts is planned – taking the effect of lot sizes into consideration – for the exact point in time they were needed. This leads to deviations in the rate of production caused by deviations on the marketing side. Kanban control cannot cope with this type of deviation because it requires constantly recurring production sequences. This forces the Kanban system to smooth out the production program for a particular period to bring about a constant demand for parts.
- MRP can be used either in job shop or in flow shop manufacturing. In principle the Kanban control can also be employed for both production systems, but in job shop manufacturing the Kanban-system leads to long transport routes and is difficult to survey. Therefore the machines should be installed where possible in line with the flow of production.
- In contrast to MRP, Kanban can also not be employed, or at least not economically, in single-order production – e.g. ship-building – or in the production of small series because in both cases there are no, or very few, repeated production sequences.
- Whereas Kanban can only be used in the production of piece-goods, MRP can be used for piece-goods and fluid-goods. It is of no consequence for MRP, when calculating the amounts required and the times they will be required, whether it is a question of parts or insert materials.
- In both systems of material flow control the parts for which each system is suitable can be determined with the help of the ABC analysis. As both systems have as an objective the minimizing of stock levels, it is worth using them for A and B materials, in other words, for materials with high and average values, as these lead to high capital costs in contrast to C materials.

- When the XYZ analysis is used it can be seen that MRP can be employed where demand sequences are steady, fluctuating and irregular. Kanban, on the other hand, can only be used where the demand is steady of fluctuating, and breaks down when the demand for the parts is irregular, because the Kanban system needs buffer stores in readiness over the whole period of time for irregularly occurring demand sequences. The strong points of MRP can be seen in this sort of situation, and not only as compared with Kanban control but also with the so-called traditional lot sizing models and stock replenishment systems.
- With MRP the allocation of work orders to the individual production centres and the control of the material flow are centralized. Centralization of material flow control is unnecessary in the Kanban system, because the production centres each control the one before and the one behind in the sense of a feedback control system. Only the construction of the feedback control loops, the first-time distribution of the Kanban cards, and the allocation of orders to the last production centre are centralized in the Kanban system. The transmission of information to each production centre, and if necessary even to the suppliers, takes place on a decentralized basis. The control of the feedback loops through the production workers leads to a decentralization not only of the flow of information but also of the disposition activities.
- The use of the computer is essential for the performance of MRP because comparatively many steps have to be carried out for the calculations and large amounts of data have to be processed. The material flow control requires data-returns from the production centres to the central control unit and this means the construction of a Production Data Acquisition system. Kanban, on the other hand, does not need, in principle, the support of EDP because the calculations are very simple and the actual control can be performed manually and is much easier to survey than an MRP control. The self-controlling feedback loops also mean that no Production Data Acquisition system is needed. A Kanban system can, however, be set up using a computer; in this case the Kanban cards can also be machine-readable.
- MRP uses the push system to control the flow of material. Kanban, on the other hand, uses the pull-system. An exception is the supply of materials purchased from outside and the deliveries from the last production centres to the finished goods stores, where the Kanban system uses the push system too.
- When planning requirements, MRP starts off with the given, average flow times and then attempts, taking capacity limits into account, to minimize the flow times in the following planning stages. Kanban attempts to influence flow times directly by organizational methods such as changes in the layout of the production facilities or adjustments to the production facilities as regards capacity. The tasks for Kanban in this area extend much further than those for MRP.
- The same thing can be said of setup times which MRP accepts as given. Kanban must necessarily influence the setup times because otherwise the small lot sizes necessary for the flexibility and for the limits on the amounts of buffer stores cannot be economically realized.
- MRP makes no direct attempt to influence quality guarantees in contrast to Kanban. MRP leaves it to the production planners to set the quality levels to be realized after considering the costs and the benefits of quality guarantees.

Kanban and MRP can be used parallel to one another. A well-known example of the simultaneous use of both systems is the so-called Synchro-System (WILDEMANN 1983, p. 4). Whether Kanban or MRP or both systems simultaneously is the solution in practice for the control of the flow of material can only be determined after an exact analysis of the areas in which each system can be used and of the preconditions for use described above. It will, however, usually be the case that Kanban control, if it can be used at all, will not be suitable for the whole spectrum of parts for all production areas in a company, and that these companies will find it more relevant to control some areas using MRP and others using Kanban.

At the moment a number of papers in the literature is especially concerned with the theoretical question, how Kanban and MRP can be compared analytically. It also seems by these studies that formal approaches to production control can be derived which include common combinations of both concepts (BUZACOTT 1988). These recent results demonstrate that further investigations are necessary to classify the relationship between Kanban and MRP.

References

Anderson, J.C., Schröder, R.G., Tupy, S.E., and White, E.M.: Material Requirements Planning, A Study of Implementation and Practice, Minneapolis 1981.

Buzacott, J.A.: Kanban and MRP Controlled Production Systems, in: Fifth International Working Seminar on Production Economics, Pre-Prints, Part 1, Igls-Innsbruck, 1988, pp. 269–306.

Dickie, H.F.: ABC Inventory Analysis Shoots for Dollars not Pennies, in: Factory Management and Maintenance, 1951, pp. 92–94.

Dinkelbach, W.: Sensitivitätsanalysen und parametrische Programmierung, Berlin-Heidelberg-New York 1969.

Eilon, S.: Elements of Production Planning and Control, New York-London-Tokyo 1962.

Fox, K.A.: MRP-II providing a natural 'hub' for Computer-Integrated Manufacturing Systems, in: Industrial Engineering, October 1984, pp. 26–31.

Kurbel, K.: Software Engineering im Produktionsbereich, Wiesbaden 1983.

Monden, Y.: What makes the Toyota Production System really tick?, in: Industrial Engineering, January 1981a, pp. 35–46.

Monden, Y.: Adaptable Kanban System helps Toyota maintain Just-in-Time-Production, in: Industrial Engineering, May 1981b, pp. 29–46.

Müller-Merbach, H.: Sensibilitätsanalyse der Losgrößenbestimmung, in: Operations Research, 1962, pp. 79–88.

Naddor, E.: Inventory Systems, New York-London-Sydney 1966.

Partridge, R.: Geringe Pufferbestände durch japanische Steuerungsverfahren, in: Fortschrittliche Betriebsführung und Industrial Engineering, 1983, pp. 374–377.

Rodat, S., and Spörl, F.: Japans Automobilindustrie, Eindrücke und Beobachtungen aus der Studienreise des REFA-AKIE nach Japan im Mai 1984, in: Fortschrittliche Betriebsführung und Industrial Engineering, 1984, pp. 323–328.

Scheer, A.W.: Wirtschafts- und Betriebsinformatik, München 1978.

Soom, E.: Die neue Produktionsphilosophie: Just-in-time-Production, 1st Part: Ein Methodenpaket zur Steigerung der Flexibilität und zur Senkung der Bestände, in: io Management-Zeitschrift, 1986a, pp. 362–365.

Soom, E.: Die neue Produktionsphilosophie: Just-in-time-Production, 2nd Part: Synchronfertigung und Kanban, in: io-Management-Zeitschrift, 1986b, pp. 446–449.

Sugimori, Y., Kusunoki, K., Cho, F., and Uchikawa, S.: Toyota Production System: Materialization of a just-in-time and respect-for-human system, in: Muramatsu, R., and Dudley, N.A. (eds.): Production and Industrial Systems, Berlin-Heidelberg-New York-Tokyo 1978, pp. 185 – 196.

Swann, D.: Execution is key to success of any systems for manufacturing material flow control, in: Industrial Engineering, October 1984, pp. 26 – 31.

Tersine, R.J.: Principles of Inventory and Materials Management, 2nd edition, New York 1982.

Wagner, H.M., and Whitin, T.: Dynamic Version of the Economic Lot Size Model, in: Management Science, 1958, pp. 89 – 96.

Wildemann, H.: Produktion auf Abruf – Werkstattsteuerung nach japanischen Kanban-Prinzipien, in: Arbeitsvorbereitung, 1983, pp. 3 – 8.

Wildemann, H.: Das JIT-Konzept als Wettbewerbsfaktor, in: Fortschrittliche Betriebsführung und Industrial Engineering, 1987, pp. 52 – 58.

Chapter 5

Effects of Call-Forward Delivery Systems on Suppliers' Serial per Unit Costs

GÜNTER FANDEL, PETER FRANÇOIS, and EUGEN MAY

I. Introduction

Improving the profitability of industrial firms involved in mass- or large serial-production by means of a more economic use of material is a topic of immediate interest. In this context modern principles of the supply of materials have been redefined, sometimes concealing well-known procurement principles which have simply been rediscovered. The two principles in question here are Just-in-Time and Kanban (WILDEMANN 1987) which have the same aim, namely to reduce uneconomic stock levels and to accelerate the lead times of the material while at the same time maintaining or imposing production flexibility as displayed by the diversity of the products on offer (SCHONBERGER 1984). The principles hold out the promise of possible reductions in stock of up to 70% and of order lead times being reduced by up to 90%, with implicitly the corresponding cost savings in the material area.

In practice, this means the principle of the supply of material synchronized with its utilization (GROCHLA and SCHÖNBOHM 1980) or stockless material supply respectively, a principle which aims at controlling the internal supply of materials and the contractual regulation of procurement procedures by means of call-forward delivery systems at the interface between the buyer and the supplier of materials (Just-in-Time). A special method of realizing Just-in-Time production is material control with the help of cards (Kanban). The sole object under consideration in this chapter will be the external procurement or supply of materials on the basis of a Just-in-Time production; this will be analysed, with particular attention being paid to the effects on costs for the buyer and the supplier.

Insofar as the procurement procedures are sufficiently standardized, i.e. the fixed order costs are so minimal that they could be ignored, the change to Just-in-Time deliveries based on an hourly production rhythm will probably always be advantageous to the buyer in the short term because he can diminish the stockkeeping costs. At the same time, however, the effects of the introduction of Just-in-Time production on the supplier's serial costs, specifically on the serial per unit costs, must be examined. If these are reduced as well, the introduction of Just-in-Time (JIT) production is profitable for both sides. If, on the other hand, JIT production causes an increase in serial per unit costs as a result of increased batch and storage costs, two cases must be differentiated. If the increase in the supplier's costs is greater than the buyer's savings, the introduction of the JIT production causes the logistic chain supplier/buyer to suffer a loss, which is economically absurd. If the increase in the supplier's serial per unit costs is less than the buyer's savings, compensations might well be in order to encourage the supplier to introduce the new procurement system, or to support it in spite of the deterioration in the cost situation. In both cases, however, the buyer will have to reckon in the long term with increased prices for the procurement lots if the supplier, for his part, is not in a position to use the potential for rationalization of JIT production.

In the following a more detailed discussion of this problem on a management basis will be found. The discussion will have practical and analytical perspectives, so that differentiating statements can be made as to the economic efficacy of the introduction of JIT production in certain, concrete decision-making situations.

II. Considerations from the Practice of the Suppliers

1. An Example of Cost Calculation from the Forge Industry

First experiences in forge firms, prime examples of suppliers of other branches, show that the introduction of JIT production makes itself felt in an increase in the batch frequency of a series and, as a result of the consequent increase in fixed batch costs, is, in certain circumstances, reflected in a considerable increase in production costs per unit of a lot with smaller lot sizes. The main cost components setup costs, start-up costs, costs of tools, administrative and sales costs, as part of fixed batch costs, along with possible extra charges for low-volume orders, which can be counted as variable batch costs, are determining factors for this development. Table 5.1 shows the changes in costs per unit

Table 5.1. Serial per unit costs caused by variations in lot sizes

Cases	1	2	3	4	5
Lot size (pieces)	100	500	1000	2000	3000
Material discount (%)	−5	−5	0	0	5
Material single costs	14.70	14.70	14.00	14.00	13.30
Rejection rate (%)	6	5	4	4	2.5
Costs of material	15.60	15.60	14.90	14.90	14.20
Production costs	11.00	10.50	10.00	10.00	9.50
Setup and start-up costs	10.00	2.00	1.00	0.50	0.33
Costs of tools	75.00	15.00	7.50	3.75	2.50
Total	111.60	43.10	33.40	29.15	26.53
Reject costs	6.70	2.11	1.34	1.17	0.66
Manufacturing costs	118.30	45.21	34.74	30.32	27.19
Admin./sales costs	11.83	4.42	3.47	3.03	2.72
Prime Costs	130.13	49.63	38.21	33.35	29.91

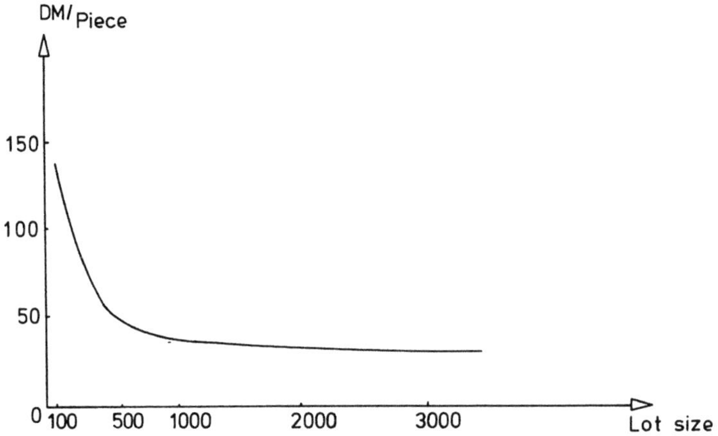

Fig. 5.1. Prime costs per unit dependent on lot size

caused by JIT production for 5 different batch sizes, based on the calculation of per unit costs of a forge firm supplying parts to the car-making industry, whereby 3000 pieces are regarded as a normal lot size. Fig. 5.1 illustrates the results of the unit cost calculations dependent on the lot size.

The graph in Fig. 5.1. shows extremely clearly that the prime costs per unit increase dramatically from 33.35 DM/unit to 38.21 DM/unit and then to 49.63 DM/unit when the lot size is halved two times from 2000 to 1000 and then to 500 units, something which would not be out of the ordinary for a supplier on the introduction of JIT production. The increase in the production costs per unit on changing from a lot size of 2000 to a lot size of 500 would be approximately 49%, and this must be examined critically in the light of possible savings for the buyer in the fields of procurement and stock-keeping. This is because optimistic prognoses usually start with the presumption that procurement and stock-keeping costs, which could be reduced by means of a more advantageous ordering principle on the lines of JIT, make up a maximum of about 40% of the prices for the procurement lots (WILDEMANN 1987). On the basis of this rough calculation the reduction in costs for the buyer caused by JIT is smaller than the increase in costs for the supplier as a result of a higher batch frequency. Clues based on practical experience for the plausibility of the increases in costs for the suppliers, where deliveries to the buyer must be carried out according to JIT, will be discussed in the following section.

2. Reasons for the Increases in Costs

Now and then reasons for the expected increases in supplier's costs on the introduction of JIT based on practical experience can be found in the literature (SCHULZ 1986). The essential points are summarized in Tables 5.2 and 5.3.

Table 5.2 shows clearly how the fixed serial costs of a medium-sized forge firm in West Germany of about 1500 DM per batch are distributed under actual

Table 5.2. Changes in the serial costs

Fixed serial costs 1500 DM (actual costs) as percentage basis.

	actual	rational-ization	JIT w/o. rat.[a]	JIT w. rat.[b]
Fixed cost factors:				
– setup costs	50%	25%	50%	25%
– start-up costs	25%	15%	25%	15%
– costs of tools	10%	10%	25%	25%
– admin. and sales costs	15%	15%	20%	20%
Extra costs as variable cost factors				
– low volume addition	–	–	30%	30%
Total	100%	65%	150%	115%
Fixed serial costs plus low volume addition	1500 DM	975 DM	2250 DM	1725 DM

[a] Just-in-Time production without rationalization.
[b] Just-in-Time production with rationalization.

Table 5.3. Parameter comparison for Japanese and German suppliers on changing to JIT production

	Japan	FRG
Number of buyers	1	10– 50
Number of different parts	12	75–300
Average series size (pieces)	15,000	2,500
Number of tools used	45	325
Change-over time per refit (minutes)	20	30
Number of the refits per month	15	80
Total change-over time (hours)	5	40
Length of contractual obligation for buyer (months)	120	1–12

conditions over the various fixed cost factors. The largest part, about 78% of the fixed batch costs, is made up of setup and start-up costs. The forge industry believes it will be able to reduce these costs to about 40% – on present fixed serial costs – in the next few years by rationalization using modern machines, and as a result of standardized methods of control. The development towards an improvement in the efficiency of actual production processes is imperative, as can be seen by the fact that without this type of rationalization the fixed serial costs will increase by 20% on the introduction of JIT (i.e. +15% for tools and +5% for administration). The considerable increase in the cost of tools is brought about by the act that smaller lot sizes mean more warming up and cooling down processes, and thus tools becoming damaged sooner, with limited possibilities only for repair. A drastic change for the worse as far as the total serial costs are concerned is seen in the costs for material procurement, the change being occasioned by the extra charges for low volume orders. These extra charges belong in principle to the variable batch costs and bring about, in effect, an increase of 30% to the present fixed serial costs. The lot would thus be 50% more expensive as far as the price factors discussed here are concerned. The increase can be partly absorbed by rationalizations in the refitting and start-up phases, so that in the end JIT coupled with a concurrent rationalization programme would mean an increase of only 15%. These calculations are based on the premise that the change to JIT for the supplier does not mean more than a quadrupling of the batch frequency of a series. It is not easy to estimate whether this is realistic in practice; everything depends on the individual case. In the reverse conclusion this supposition, together with the data from Table 5.1, would mean that a higher batch frequency would lead to a considerable worsening of the results shown in the columns 3 and 4 in Table 5.2. This premise is certainly supported in part by the impression gathered from Table 5.3, in which a comparison is made of the combination of parameters relevant to the introduction of JIT for Japanese and German suppliers in the forge industry. German suppliers have to cope with a much more complicated buyers' structure and a wider parts spectrum, both of which combine to make more than 5 times as many refits per month necessary, thus causing the starting conditions for JIT to be comparatively worse than in Japan, and also making the experience gathered there not directly relevant to German conditions. A further serious difference between the two countries, that of the long-term contractual binding of the buyer to the supplier, as far as amounts to be bought are concerned, will be gone into later.

The empirical results of the effects of call-forward systems or JIT production on the supplier's serial costs will now be followed by some analytical considerations directed towards showing possible cost developments for buyers and suppliers which can be expected by reason of different cost combinations, and which can be used to assess the total effect of the introduction of JIT production.

III. Analytical Considerations on the Introduction of Just-in-Time Production

1. The Buyer's Motives

For some time now competition in many areas has become considerably harder, one of many reasons being the effects of stagnation and product saturation. Producers have, therefore, been searching for ways and means of increasing their competitiveness in order, at least, to retain their market position. Just-in-Time production, with as much stockless material supply as possible, seems to be a promising concept, and is one which has been practised for years by the Japanese competitors, especially car manufacturers (MONDEN 1981 a, 1981 b). JIT involves a changeover to smaller order volumes, which would lead to higher procurement costs and lower stock-keeping costs, if the previous optimum order volumes were

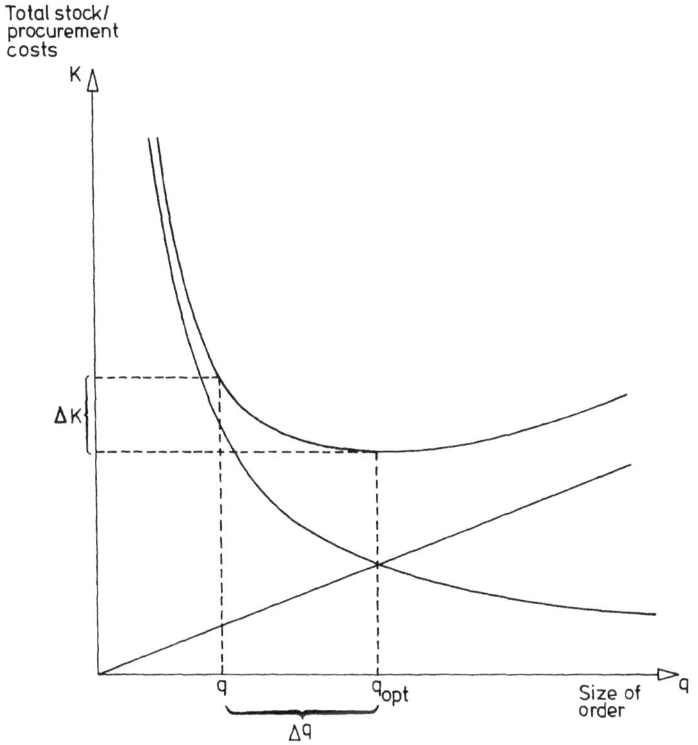

Fig. 5.2. Total costs dependent on size of order

used as a basis, but on the whole would lead to higher total costs for procurement and stock-keeping. Fig. 5.2 illustrates the connection between the amount ordered and the total costs (KILGER 1973; TERSINE 1982). In order, therefore, to make the changeover to JIT appear reasonable to the buyer, along with the aims mentioned above, the following reasons must be alternatively present:

(1) Fixed Order Costs Can Be Reduced.
Fixed order costs are procurement costs which are dependent on the frequency of the orders but not on the volume. One possibility of reducing these fixed order costs is the employment of data processing in the procurement department. For example, a production planning system could generate an order recommendation following an analysis of requirements. If the stock clerk agrees with the proposed order an order form is automatically filled in and delivered to the supplier via remote data transmission. The costs of determining the amounts to be ordered, of drawing up the order forms, and of transmitting them to the supplier can be considerably reduced in this way. A similar effect could be achieved by making the supplier responsible for quality controls. If faults are found the supplier is made liable, whereby penalties and liability claims are regulated in the contract. Passing on responsibility for the goods to the supplier would, however, only be reasonable in

particular cases, for example, when the product does not have to satisfy stringent safety regulations, or is not made up of several smaller parts, or does not have a high value.

(2) Stock-Keeping Costs Are Higher Than Orginally Expected.
This can happen, for example, when the rate for stock-keeping costs is set too low, because insufficient allowance was made for capital lockup costs, insurance, and the tax on stocks. The effect of these types of stock-keeping costs dependent on the value of the stock is not unimportant to the success of the firm, as can be seen from an analysis (WÄSCHER 1987) which showed that capital tied up in stock in the mechanical engineering industry totalled 39% of the balance-sheet value in 1981.

(3) A Combination of (1) and (2).

If one concentrates without limitation of the universality on a development along the lines of (1) for a serial product on the condition that the buyer's fixed order costs can be permanently reduced, and if it is further supposed that the buyer can determine his optimal ordering policy on the basis of the Harris formula with an infinitely rapid addition to stocks (HARRIS 1915) and, because of his market position, is able to force the supplier to accept this policy, the steps shown in Table 5.4 may illustrate the gradual changeover of the buyer to a JIT ordering policy in order to save costs.

Table 5.4. Decision-making and cost situation for the buyer on change to JIT ordering

Step	1	2	3	...	S
Planning horizon T (weeks)	16	16	16	...	16
Total requirements x (pieces)	3200	3200	3200	...	3200
Fixed order costs c (DM/order)	640	160	40	...	10/49
Stock-keeping cost rate l (DM/piece × week)	0.10	0.10	0.10	...	0.10
Optimal order size[a] q (pieces)	1600	800	400	...	200/7
Order frequency h	2	4	8	...	112
Order costs $K_B = c \cdot h$	1280	640	320	...	22.86
Stock costs $K_L = q/2 \cdot l \cdot T$	1280	640	320	...	22.86
Total costs $K = K_B + K_L$	2560	1280	640	...	45.72
Savings in costs $-\Delta K^A$		+1280	+1920	...	+2514.28

[a] $q = \sqrt{\frac{2 \cdot c \cdot x}{l \cdot T}}$

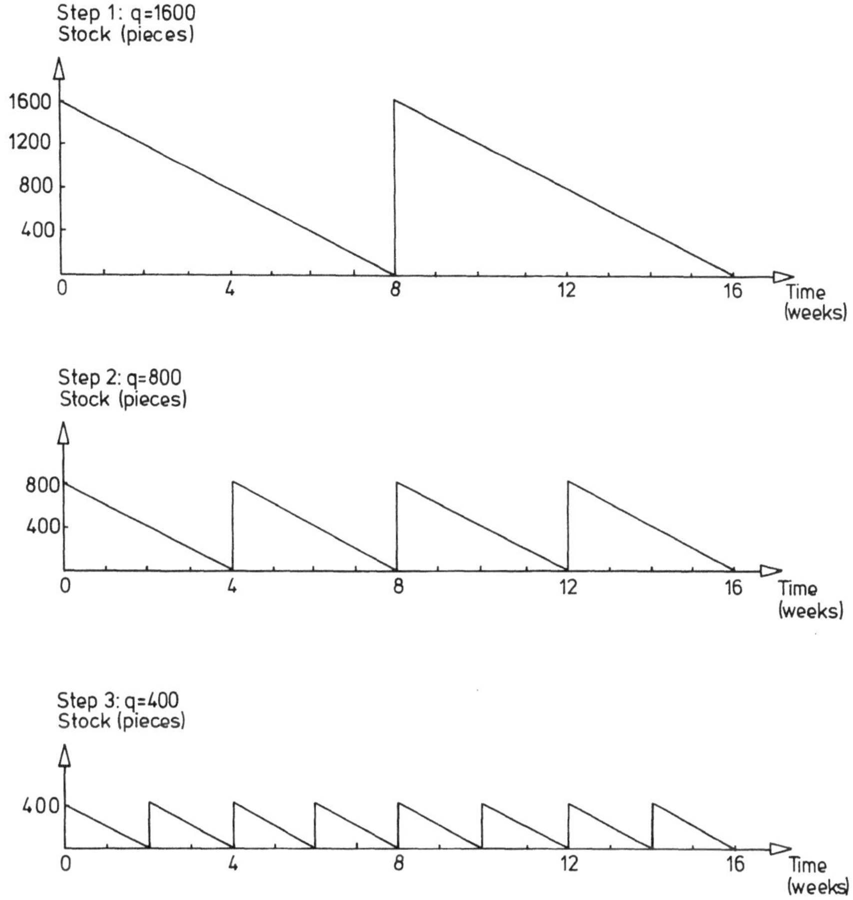

Fig. 5.3. Stock processes for the buyer

In the starting situation (step 1) the planning horizon equals $T = 16$ weeks and the total demand is $x = 3200$ pieces. These data are retained for the further distinctions of cases. The fixed order costs then develop in such a way that the optimum order size, starting with $q = 1600$, is halved twice and in the end is 200/7, which leads gradually to daily ordering, i.e. $h = 112$, in comparison with the original order frequency $h = 2$. Given this sort of alteration on costs the changeover to JIT production is always worthwhile for the buyer. Fig. 5.3 shows the graph of the development of stock-keeping processes resulting from the changeover to JIT for the buyer.

It remains to be said that the order size has a lower limit dependent on transportation conditions. From an economical point of view it would be sub-optimal, for example, to use half-loaded trucks, containers, pallets etc. for transportation.

The question which remains at the end of the survey of the situation for the buyer is: What is the cost situation for a serial product as far as the supplier is concerned, and how big is the total effect on the logistic chain supplier/buyer? An examination of these aspects will take place in the following section.

2. Consequences for the Supplier

(a) Possible Starting Situations for the Supplier

If the introduction of JIT production is based on the initiative from the buyer, and if his position in the market enables him to force the initiative through, the supplier has no alternative but to accept the new delivery dates and order sizes for the serial products. Developments in the areas of serial batch costs and stock-keeping costs on a gradual changeover to JIT production are now of interest for various combinations of the supplier's costs. Three different combinations must be differentiated and are described here as A, B, and C. The relevant data are shown in Table 5.5.

The fixed delivery dates and volumes mean that in general the suppositions of the Harris model are not fulfilled as far as the supplier is concerned, so that the lot size formula cannot be used to determine an optimal batch policy. On the contrary, the possibilities for action which a supplier has – choosing the batch frequency – must be examined separately as to the effects on costs, and then the optima must be figured out using the principle of complete enumeration.

Table 5.5. Data for different combinations of the supplier's costs

Supplier	A	B	C
Planning horizon T (weeks)	16	16	16
Total production x (pieces)	3200	3200	3200
Production speed z (pieces/week)	400	400	400
Fixed batch costs a (DM/batch)	100	300	800
Stock-keeping cost rate l' (DM/piece × week)	0.08	0.08	0.08

(b) Development Trends on the Introduction of JIT Production

If we start with step 1 – i.e. the starting situation – for the buyer, with the optimal order size $q = 1600$ and therefore the order frequency $h = 2$, the supplier has the choice of one or two batches, and this leads, given current production and delivery conditions, to the stock flows shown in Fig. 5.4. As can be seen from both parts of Fig. 5.4, the supplier must start production at the end of the forth month at the latest to avoid shortages. An earlier start would lead to an unnecessary increase in stock costs, and would, therefore, not be efficient. The two

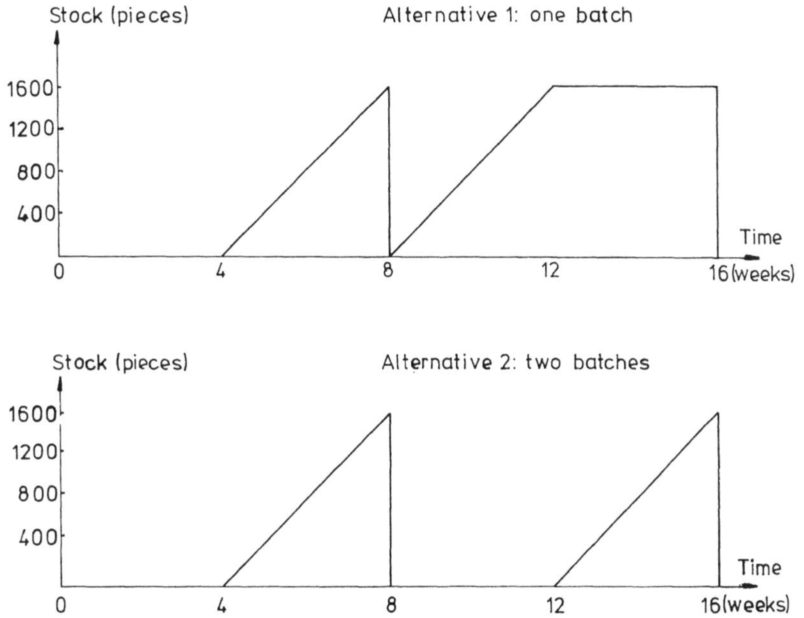

Fig. 5.4. Alternatives for the suppliers given $q = 1600$ (step 1 for the buyer)

Table 5.6. Suppliers' costs for buyer's step 1 (starting situation)

Supplier	A	B	C
Alternative 1: one batch	1124	1324	1824*
Alternative 2: two batches	712*	1112*	2112

alternatives illustrated in Fig. 5.4 lead to total costs for suppliers A, B, and C as shown in Table 5.6. The optima are indicated by asterisks. It can be seen that, given a twice repeated delivery of $q = 1600$, minimal costs for suppliers A and B result from two batches $q' = 1600$, and for supplier C from one batch $q' = 3200$.

Step 2 for the buyer, with an order size of $q = 800$ every four months – a total of four deliveries –, means that the suppliers have the opportunity of executing the total production $x = 3200$ in one, two, three, or four batches. In several of these alternatives, however, there are subcases which have to be differentiated according to the distribution of the batches over the 16 weeks. The alternatives and their subcases are illustrated in Fig. 5.5 as far as the related stock processes are concerned. In alternative 2, the stock processes in the subcases 1 and 2 are obviously identical, but are dominated in respect of the costs by subcase 3 because of the fixed batch costs. In alternative 3 two further distributions of the three batches are possible which, apart from shiftings, lead, in principle, to the same stock processes and to the same costs.

Table 5.7 shows the suppliers' total stock-keeping and batch costs resulting from these alternatives.

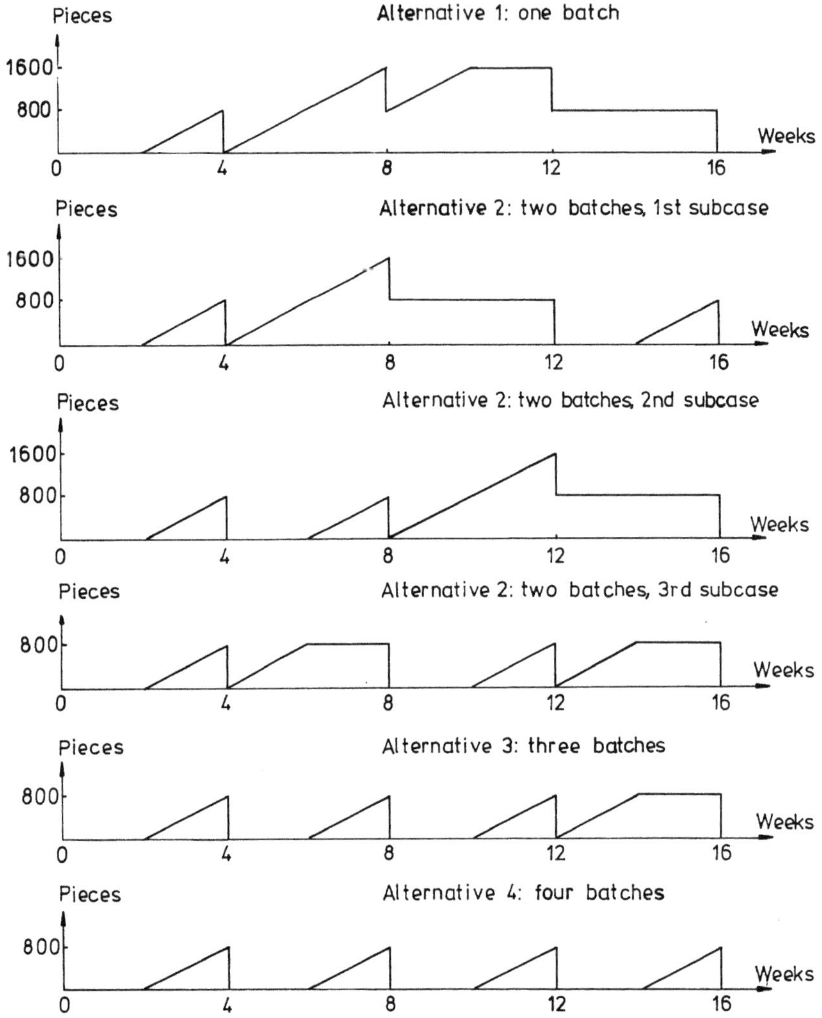

Fig. 5.5. Alternatives for the suppliers given $q = 800$ (step 2 for the buyer)

If Table 5.7 is compared with Table 5.6 it can be seen that it would pay supplier A to move towards the buyer's half-sized order because he now has four batches with minimal costs because of the relatively low fixed batch costs. On the other hand, the batch policy of the starting situation remains optimal for suppliers B and C. Their costs do not change. In general, this can be proved for all suppliers from the parity of costs of the alternatives 1 and 2 in the starting situation (step 1) and the subsequent situation (step 2) as can clearly be seen in Fig. 5.6.

It can be seen that the upper and lower stock processes are each identical as to costs, because the batch frequencies and average stocks are the same. The latter can be controlled using the parity of the surface areas.

Table 5.7. Costs to the suppliers for buyer's step 2

Supplier	A	B	C
Alternative 1: one batch	1124	1324	1824*
Alternative 2: two batches	712	1112*	2112
Alternative 3: three batches	684	1284	2784
Alternative 4: four batches	656*	1456	3456

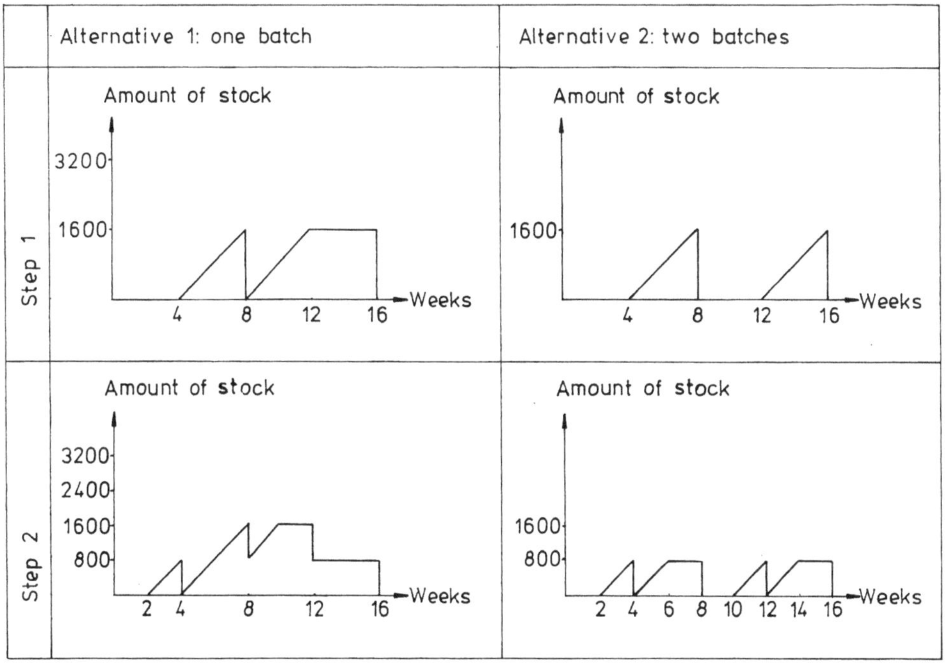

Fig. 5.6. Parity of costs in starting and subsequent situation

Similar considerations have to be set in motion for step 3 with the order size $q = 400$, whereby 8 alternatives for the suppliers with a total of 22 subcases must be gone over in order to find the optimal batch policies for the suppliers. Table 5.8 shows the costs of the alternatives, leaving out the subcases which are not relevant as to costs.

What is most conspicuous is that for supplier A three batches are now optimal instead of four batches as before. This is because of the fact that stock usage takes place twice as often as before and therefore stock costs are lower. For suppliers B and C nothing has changed as to the optimal state of the existing batch frequency, or as to their minimum costs. It can be further seen that batches over and above the optimal batch frequency cause the suppliers' batch and stock-keeping costs to increase monotonically.

In step S (daily delivery) the start of production for the suppliers draws so near to the point in time $t = 0$ and the stock processes become so connected with

Table 5.8. Costs to the suppliers for buyer's step 3

Supplier	A	B	C
Alternative 1: one batch	1124	1324	1824*
Alternative 2: two batches	712	1112*	2212
Alternative 3: three batches	652*	1252	2752
Alternative 4: four batches	656	1456	3456
Alternative 5: five batches	724	1724	4224
Alternative 6: six batches	792	1992	4992
Alternative 7: seven batches	844	2244	5744
Alternative 8: eight batches	928	2528	6528

Table 5.9. Optimal batch policy and costs to the suppliers for step *S*

Supplier	A	B	C
Optimal lot size q	1000	1732.05	2828.43
Batch frequency h	3.2	1.85	1.13
Total costs K^z	640	1108.52	1810.20

one another that the suppliers' batch policy for the planning horizon $T = 16$ weeks which is held to, can be approximately determined by means of the Harris model with successive stock additions and continuous stock usage (KILGER 1973; TERSINE 1982). The stock additions rate reads $z = 400/7$, the stock usage rate $y = 200/7$, and the net stock additions rate reads $z - y = 200/7$. The results are shown in Table 5.9. It can be seen that, apart from the deviations caused by the approximation, the same optimal batch policies exist for the suppliers as in step 3. It must be pointed out here that the total costs K^z are only approximate costs, unlike the costs in the above calculations. The real total costs are higher than the optimal costs calculated by means of this method.

The cost developments for the suppliers and the buyer in the case of a gradual changeover to the JIT principle, as derived from the analytical considerations, are shown in Table 5.10.

Beginning with the starting situation (step 1) the changeover to JIT production – on the assumptions made in Section III.1. – always pays for the buyer as far as costs are concerned. The changeover can also pay for the supplier (supplier A); the position of the suppliers can in no way deteriorate when the material requirements and the planning horizon of the starting situation are kept to, and they are therefore free to plan their own optimal batch policy. It is, however, exactly this implicit assumption which contains the crux. If, for example, the buyer, for reasons of flexibility, decides on changing over to the JIT principle not to commit himself to a total requirement of 3200 pieces for $T = 16$ weeks, but commits himself simply on an informatory basis to an analogical total requirement of 400 pieces for the shorter planning period of $T = 2$ weeks – and this is more or less what happens in practice – it only remains for supplier C as in step

Table 5.10. Development of costs for suppliers and buyer and total effect of changeover to JIT production

Supplier	A	B	C
Step 1 (starting situation)			
Costs for suppliers	712	1112	1824
Costs of buyer	2560	2560	2560
Step 2			
Costs for suppliers	656	1112	1824
– Savings against step 1	56	0	0
Costs for buyer	1280	1280	1280
– Savings against step 1	1280	1280	1280
Total savings	1336	1280	1280
Step 3			
Costs for suppliers	652	1112	1824
– Savings against step 1	60	0	0
Costs for buyer	640	640	640
– Savings against step 1	1920	1920	1920
Total savings	1980	1920	1920
Step *S*			
Costs for suppliers	640	1108.52	1810.20
– Savings against step 1	72	3.48	13.80
Costs for buyer	45.72	45.72	45.72
– Savings against step 1	2514.28	2514.28	2514.28
Total savings	2586.28	2517.76	2528.08

3 to bring the series out 8 times if he wants to avoid the risk of wastage. This increases his costs – see Table 5.8 – to 6528 DM, which means an increase of 4704 DM compared with the starting situation. These figures are also valid when it is assumed that stock usage takes place continuously over the 2 weeks. Thus the maximum savings in costs for the buyer on a complete changeover to JIT production of 2514.28 DM are more than overcompensated. The introduction of the JIT principle would not make economic sense because the logistic chain supplier C/buyer would deteriorate considerably as far as costs are concerned. In order to support the argument against a reduction in the reported requirements and planning periods the suppliers claim in addition that material procurement can take up to 3 months, and tool replacements can take from 7 to 10 weeks. This is in contrast to the Japanese findings, as can be seen from Table 5.3. The call from the suppliers for specifications for material requirements to be given by the buyer on a target basis over and above the shortened planning period is, therefore, understandable.

A further point in the interpretation of the findings should be noted. The analytical arguments have neglected the problem of the cost of transporting material from the supplier to the buyer. If the supplier has to bear the costs his position deteriorates still further compared with the situation described here. If the buyer bears the costs, his saving in costs is lower than those shown here.

3. Deriving a Formula for Determining the Optimal Lot Size of a Serial Product with Just-in-Time Deliveries

Up to now the optimal batch policy for a serial product has been determined for the suppliers by complete enumeration, in other words, the costs of all possible batch combinations have been compared with one another. It is interesting to note that it can be optimal to produce lots of different sizes within a planning horizon. This can be seen from Table 5.8 for supplier A and from the corresponding graph of the stock process in Fig. 5.7. The observations here, however, with equal amounts of material required, seem to point to constant lot sizes. The reason for this contradiction is the restriction to a fixed planning horizon. Thus it is impossible to produce n equally large lots in a planning horizon when the time between the resetting of the lots is longer than the length of the planning horizon divided by n. Fixed planning periods lead, by way of lots of different sizes, to suboptimal results. This can be shown by joining several planning horizons together. For this purpose it is proposed to start from the optimal batch combinations in planning period T as shown in Fig. 5.7.

Lots of different sizes are produced in the given planning period. If, however, the planning period is repeated so often that the sum of the units produced in the smaller lots equals the production of one or more larger lots, the following cases must be distinguished:

1) Substituting one or more large lots for the small lots brings about a cost advantage.
2) Substituting small lots for the large lots brings about a cost advantage.
3) Substitution has no effect on costs (indifference).

In the first two cases lots of different sizes are always dominated as to costs by lots of the same size. In case 3 different lot sizes can be optimal within a planning period, but they can never dominate lots of constant size. Using Fig. 5.7 as an example, if the planning period is tripled, 3 small lots can be replaced by 2 large lots, or vice-versa. This leads to reductions in costs if 3 small lots and 2 large lots differ from one another as to costs. If not, the substitution does not bring any change in costs (indifference). For this reason lots of different sizes can be left out of the considerations for the time being, if the premise of a fixed planning period is abandoned. A solution is then found of either an optimal

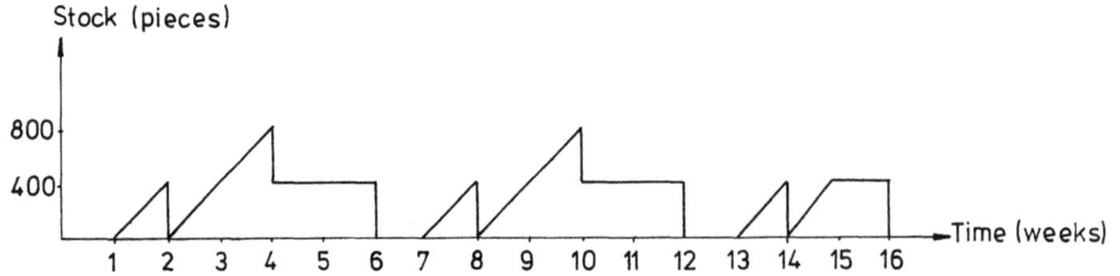

Fig. 5.7. Stock process for supplier A with optimal batch combination in step 3 and restriction to a fixed planning period

batch policy or, in the case of cost parity, two optimal batch policies with lots of equal sizes. In the special case of indifference optimal batch policies with lots of different sizes can be derived from both batch policies with equal sizes by interchanging large and small lots in corresponding relationships: In the example chosen, 3 small lots are replaced by 2 large lots, or 2 large lots exchanged for 3 small lots. The costs which accrue to the supplier when lot sizes remain constant must now be examined.

Stocks costs are dependent on the rate for stock costs and the stock process. The accumulated stock $\bar{L}$ from lots of different sizes can be determined for Just-in-Time delivery by using a formula. To arrive at the formula the following symbols are used:

$\bar{x}$ the quantity reached by equal distribution of the total requirements x over the delivery dates.

n number of the quantities $\bar{x}$ produced in a lot. Multiplying n with $\bar{x}$ gives the size of the lot.

T length of planning period.

t length of the period from the beginning of the planning horizon to the first delivery date or between two delivery dates.

z speed of supplier's production in quantity units per time unit.

a fixed batch costs in money units per batch.

l' stock cost rate in money units per units of time and quantity.

The accumulated stock on hand will now be shown using a graph (see Fig. 5.8) for the cases $n = 1$ to $n = 5$ in order to make the inductive derivation of the formula clear.

Using this as a basis, the accumulated stock on hand $\bar{L}(n)$ for arbitrary n can be determined as follows:

$$\begin{aligned}\bar{L}(n) &= n\left[\frac{\bar{x}^2}{2z}\right] + \bar{x}\sum_{k=0}^{n-1} k\left[t - \frac{\bar{x}}{z}\right] \\ &= n\left[\frac{\bar{x}^2}{2z}\right] + \frac{n}{2}\cdot(n-1)\cdot\bar{x}\cdot\left[t - \frac{\bar{x}}{z}\right] \\ &= \frac{n\cdot\bar{x}}{2z}\left[(n-1)\,tz - (n-2)\cdot\bar{x}\right]\ .\end{aligned}$$

It is not intended to prove the formula here using complete induction.

In order to provide a basis for comparison for the costs of different large-sized lots the unit costs $k(n)$ of a lot will be calculated. To do this, the total costs of the lot are divided by the lot size, which is determined by $n \cdot \bar{x}$. One receives:

$$\begin{aligned}k(n) &= \frac{a + l'\cdot\bar{L}(n)}{n\cdot\bar{x}} \\ &= \frac{a}{n\cdot\bar{x}} + \frac{l'}{2z}\left[(n-1)\,tz - (n-2)\cdot\bar{x}\right]\ .\end{aligned}$$

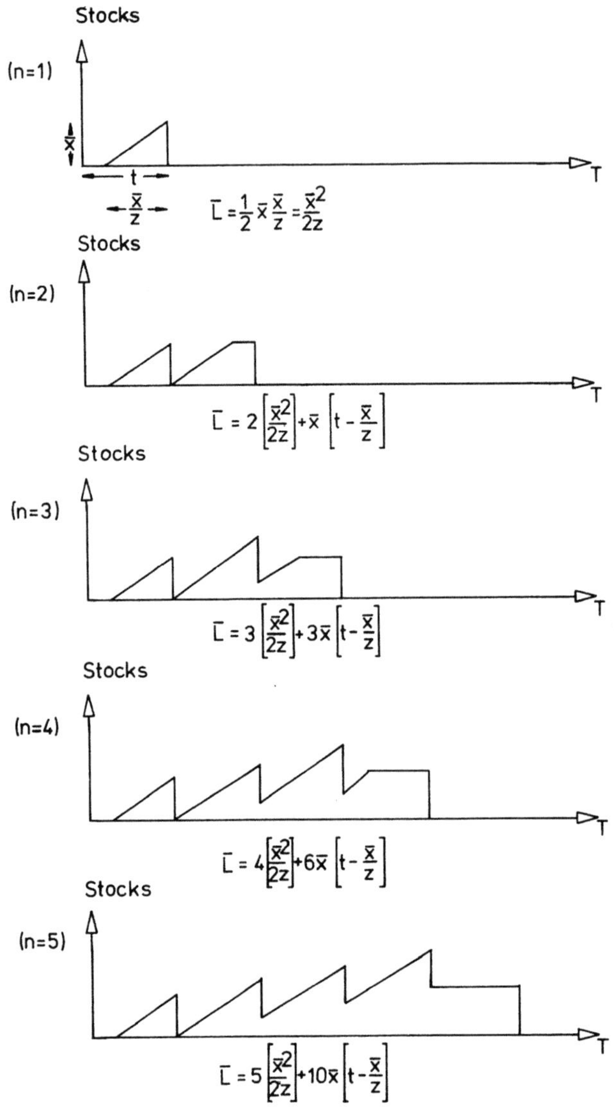

Fig. 5.8. Accumulated stocks for $n = 1$ to $n = 5$

The optimal n^* is the one which minimizes the costs per unit of the lot and therefore the serial per unit costs dependent on n, given regularly recurring batch patterns. If n^* is multiplied by $\bar{x}$ the optimal lot size q^* is found

$$k(n^*) = \min_n \{k(n)\} \ ,$$
$$q^* = \bar{x}n^* \ .$$

The application of this formula to suppliers A, B, and C leads to the results shown in Table 5.11.

Table 5.11. Costs of step 3 for suppliers A, B, and C with an open planning horizon

Supplier	A	B	C
n^*	3	4	7
Batch frequency in T	2.67	2	1.14
Lot per unit costs $k(n^*)$	0.2033	0.3475	0.5657
Total costs per planning horizon	650.67	1112	1810.24

The formula developed here determines a lot size which leads to minimum total costs in the subsequent periods, if delivery rates and amounts are fixed. The savings in costs which occur in comparison with complete enumeration can be seen by contrasting the optimal total costs in Tables 5.8 and 5.11. A further advantage over complete enumeration is that considerably fewer calculations are necessary to determine the optimal lot size.

The formula also has the advantage for the supplier over the modified Harris formula of a better cost formation. Whereas the modified Harris formula cannot generally (or at least can only approximately) determine the optimum lot size, and therefore the costs, the formula enables actual costs to be calculated. The application of the modified Harris formula proceeds approximately on the basis of a finite stock usage speed and thus transforms the discontinuous stock usage of JIT deliveries (with regular, successive delivery dates) with infinite speed into a continuous stock usage with finite speed. If the number of delivery dates is moving towards the infinite, the modified Harris formula will always move nearer to the actual values, because the discontinuous approach converges to the continuous. The Harris formula can only be used in these cases, if at all. In reality the supplier will always be faced with a stock process different from the Harris formula's approximately determined process because of the fixed delivery quantities and set delivery dates. The discrepancy gets bigger the earlier the approximation is used for one of the steps 1 to S. For example, if the total costs on the basis of the actual process for step 3 are calculated, supplier A is faced with 11% higher costs than with the approximation according to HARRIS. The costs based on the Harris formula are (for step S as well) 640 DM and the actual costs are 710.40 DM per planning period. In contrast to these actual costs the formula set out above leads to costs of 650.67 DM per planning period for supplier A in step 3 (see Table 5.11) and therefore means savings in costs of 8.41% as against the application of the modified Harris formula.

References

Andler, U.: Rationalisierung der Fabrikation und optimale Losgröße, München 1929.

Grochla, E., and Schönbohm, P.: Beschaffung in der Unternehmung, Stuttgart 1980.

Harris, F.: Operations and Costs, in: Factory Management Series, Chicago 1915, pp. 48 – 52.

Kilger, W.: Optimale Produktions- und Absatzplanung, Opladen 1973.

Monden, Y.: What makes the Toyota Production System really tick?, in: Industrial Engineering, January 1981a, pp. 35 – 46.

Monden, Y.: Adaptable Kanban System helps Toyota maintain Just-in-Time-Production, in: Industrial Engineering, May 1981b, pp. 29 – 46.

Pack, L.: Optimale Bestellmenge und optimale Losgröße, Zu einigen Problemen ihrer Ermittlung, Wiesbaden 1964.

Schonberger, R.J.: Just-in-Time Production Systems: Replacing Complexity with Simplicity in Manufacturing Management, in: Industrial Engineering, October 1984, pp. 52 – 63.

Schulz, J.: Disposition von Zulieferteilen der Automobilindustrie in Bandbreiten, in: VDI-Z, June 1986, pp. 49 – 52.

Tersine, R.J.: Principles of Inventory and Materials Management, 2nd edition, New York 1982.

Wäscher, D.: Gemeinkosten-Management im Material- und Logistik-Bereich, in: Zeitschrift für Betriebswirtschaft, 1987, pp. 297 – 315.

Wildemann, H.: Just-in-Time Production, 2nd edition, München 1987.

Chapter 6

Regulation of the Purchase of Pipe- and Line-Based Sources of Energy in Industrial Companies

JOHANNES WOLF

I. Problem and Approach

Up until the beginning of the seventies in general only secondary importance was attached to energy as a production factor in industrial companies. This assessment underwent a change as a result of some radical economic and political events, in particular on account of the crude oil price crises in 1973/74 and 1979/80. This fact has surprisingly still hardly been taken into account by economic research which means that companies do not have sufficient decision concepts of economic energy utilization at their disposal and a whole series of questions therefore need to be examined.

This paper deals with a problem of energy orientated production process planning, namely the regulation of the purchase of pipe- and line-based sources of energy – gas and electricity. This concerns an area of responsibility the significance of which has been widely recognized in the literature and to which a good deal of attention is devoted in company practice (HUGEL and SCHMITZ 1977, p. 75; KERN 1981; KERN 1984; KILGER 1981, p. 387; RICHARTS 1976; VERBAND DER ENERGIEABNEHMER 1978). The subject of the following expositions are various' types of quantitative solution contributions for this complex of questions.

The practical significance of regulation of the purchase of pipe- and line-based sources of energy stems from price settlements of the public utility organizations according to which the energy costs are increased on account of short-term purchase peaks (FINSINGER and KLEINDORFER 1981; KILGER 1981, p. 386; VERBAND DER ENERGIEABNEHMER 1978). On the basis of the published contributions to this problem area and thanks to personal contact to company practice it has been largely possible to present case studies and to demonstrate various solution methods with them.

The main regulation forms which come into question entail regulation by means of purchase interruptions and by the establishment of machine schedules. The structure of this paper has been subdivided accordingly. Whilst a number of articles have appeared in the literature about the concept of purchase interruptions (FATTI 1983; FATTI 1984; VOSS and WARTMANN 1977), hardly any attention has been paid to machine utilization programmes in the context which is of interest here (a first attempt on this concept is to be found in WOLOBRINSKI and SORIN (1976). For the latter method a decision model is to be formulated on the basis of the mixed integer linear programming and the way in which it can be applied will be demonstrated with the help of exemplary calculations.

II. Regulation of the Demand for Pipe- and Line-Based Sources of Energy with the Help of Purchase Interruptions

Within the framework of the regulation of the demand for pipe- and line-based sources of energy the first aspect to be dealt with here is the modus operandi of purchase interruptions on the part of the consumer: The temporary partial or total stoppage of the withdrawal of gas, electricity, or district heat from the network of the supplier should bring about a smoothing of the purchase of energy.

The first case of this form of regulation which is to be dealt with here is that of gas supplies with the possibility of storing larger quantities of this energy source within the firm. The subsequent expositions then refer to regulation measures in cases in which electricity is used. The other pipe- and line-based energy source, district heat, for which the price regulations are similar to those for gas and electricity but which overall is used relatively little, is not to be separately examined.

1. Purchase Interruptions in the Utilization of Gas with Storage Possibilities

A South African steel plant, in which gas is used for production purposes, is to serve as the reference company of the case study which is now to be looked at (Fatti 1983; Fatti 1984). It is possible to install storage tanks in which the gas, for which there is no current demand from the production, can be stored. The incentive behind smoothing the consumption is to be found in the fact that the peak purchases per hour and per day measured within one year in comparison to the average hourly and daily purchases determine the work price of a half-yearly accounting period. This price regulation means that a single purchase quantity which is above average can result in marked price increases.

In the face of these facts an attempt has been made to simultaneously lay down the internal upper limit of gas purchased per day and per hour respectively and to set up the storage capacity in such a way that the overall supply costs are minimized. Once the limits are reached it has to be ensured, on the one hand, that the gas supply from the utility company is temporarily totally stopped and, on the other hand, that stored gas, in as far as it is available, is utilized.

The alteration in the action parameters has contrary effects on the individual cost components. The higher the upper purchase limits are set the higher are the gas costs in the case of its exhaustion but the lower, by comparison, are the costs caused by production outages – until they have sunk to zero. Secondly, if the installed storage capacity is enlarged then the investment and other storage costs increase accordingly, whilst the costs arising on account of production outages are in turn reduced.

In the analytical method employed for the minimization of the expected value of the total gas supply costs for one day the quantities of gas needed per hour and per day respectively in order to carry out the production are integrated as empirically obtained distributions. A conceivable course of the demand for gas and the meeting of this demand, after the limits for the purchase stoppages have been laid down and after the installation of the storage units, is exemplified for one day in Fig. 6.1.

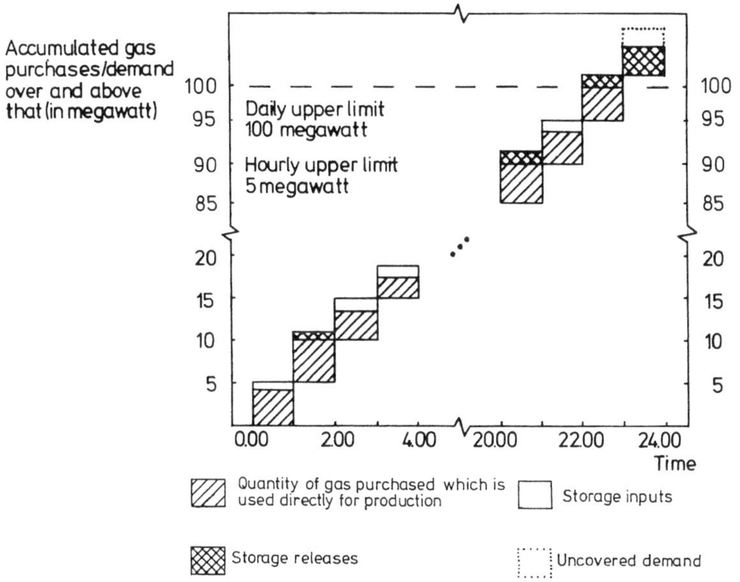

Fig. 6.1. Gas demand and the way in which it is met in the case of regulation by means of purchase interruptions and utilization of stored gas

For the fixed upper limits of 5 megawatts per hour and 100 megawatts per day Fig. 6.1 illustrates the quantity of gas which was immediately used for production, the storage inputs and releases, and the demand which could not be met because no supply quantities or stored quantities were available. If at a certain point in time the gas storage tanks are full on account of the preceding developments in demand and supplies, then the highest possible hourly purchase might not be exhausted in the interval in question – in this example this occurs between 3 and 4 o'clock. Consumption stoppages are carried out in those cases in which the hourly and daily upper limit would otherwise have been exceeded by the respective quantity of gas used. In the latter situation this means that the hourly purchases accumulated throughout the whole day reach the daily limit and that over and above this there is still a demand. In the example presented here at the end of the day there is also a demand which cannot be met with the result that there is a production failure because there is not sufficient stored gas to meet the additional demand.

2. Switch-Off Strategy in the Case of Electricity Utilization

The storage of energy, which in the case of gas is basically possible without any difficulties, raises serious problems in the case of electricity and consequently for this source of energy this method has up until now not found its way into industrial production to any great extent. Instead of first undertaking regulatory measures when the limit has been reached and then bridging the purchase bottlenecks which arise with the help of the stored fuel – as in the case which has just been described – it is also possible, when it looks as though the purchase limit is going to be exceeded, to prophylactically switch off certain consumers, a method which in the case of electricity is frequently employed.

The choice of the equipment which, if necessary, has to be switched off is made in such a way that the disruptions in the production process, as a result of possible partial interruptions in supply, are kept to a minimum. The determining of the situation in which at any one time the purchase limit is regarded as threatened and which thus causes the switch-off of equipment can on this occasion be effected by means of explicitly calculated consumption prognostications which attempt to take the respective facts of the individual case into consideration. It can, however, also be carried out without such a prognostication on the basis of standardized assumptions.

(a) Switch-Off of Equipment without Explicit Consumption Prognostications

The presentation of a case study of a switch-off strategy without explicit consumption prognostications can be effected here with reference to a large-scale German mechanical engineering firm which was kind enough to make the relevant information available. In a subsidiary plant a maximum monitoring system has been installed in order to keep the electricity costs resulting from purchase spikes as low as possible. A power consumption price agreement has been made with the energy supply company in accordance with which the highest 15 minute average power consumption during every month of one year is determined. The annual power costs are then calculated according to the average of the three highest of these twelve 15 minute values. The regulation of purchase is seasonally independent and can therefore not be limited to the winter months because the electricity is utilized exclusively for the production process and not for heating the buildings.

The main problem of this regulation method, as far as the plant's management is concerned, is finding electricity consumers which can be temporarily switched off without causing disruptions. Installations are preferred for this purpose which are able to store the intermediate sources of energy or useful forms of energy which are won from electricity like, for example, compressors with compressed air reservoirs or heating devices for galvanization equipment.

The maximum power consumption allowed on average every 15 minutes is programmed into the monitoring system, which in the case at hand amounts to 7.5 megawatts – whereas a typical daily consumption curve before the installation of the maximum monitoring system revealed quarter hourly mean values of up to 7.9 megawatts. Within each of these quarter hourly intervals the measuring and control system continually establishes the actual average power consumption and compares it with the limit which has to be kept, with a view to intervening in the purchase of electricity if necessary. Accordingly an explicit electricity consumption prognostication is not undertaken up to the end of the respective quarter of an hour. Fig. 6.2 illustrates this approach.

If the average power consumption reaches the currently operative value of the switch-off limit then the monitoring system interrupts the electricity supply to

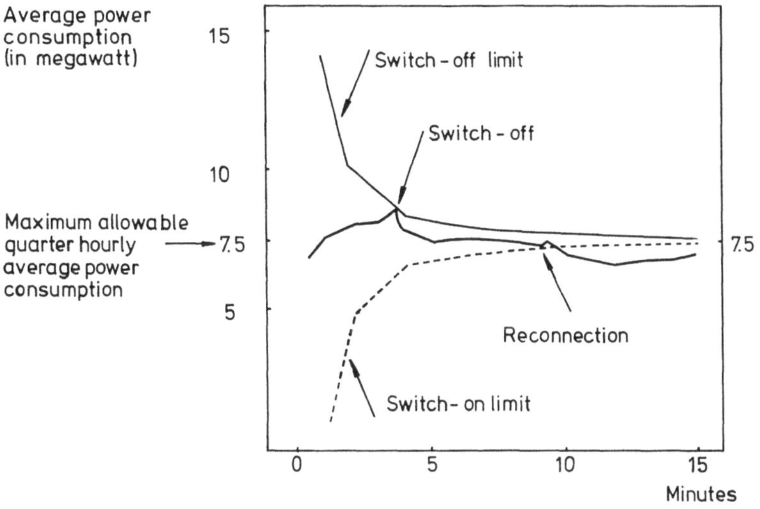

Fig. 6.2. Average power consumption in the case of the regulation of the purchase of electricity by means of a switch-off strategy without explicit consumption prognostications

consumers in accordance with prescribed priorities. If the lower limit line is subsequently reached then the relevant consumers are turned on again in the reverse order.

At the beginning of a quarter hourly interval it is permitted to exceed the maximum average power consumption of 7.5 megawatts to a greater extent without a switch-off being triggered off so that the possibility of „self-regulation“ during the rest of the interval is retained. Towards the end of each 15 minute period it is however advisable to be increasingly cautious in this respect so that the switch-off limit is lowered more and more until it eventually reaches the maximum allowable mean quarter hourly power consumption. On the other hand the switch-on level correspondingly approaches this point from below. The gradient of the two lines which serve to initiate interventions can be varied within certain limitations in accordance with the intention of the decision-maker.

The monitoring system records which consumers were switched off and for how long in order to keep below the upper limit. From this documentation it can then, for example, be seen a posteriori whether the maximum allowable average power consumption would in fact have been exceeded without intervention and if so to what extent.

In the case at hand the investment in the maximum monitoring system was in retrospect assessed as being advantageous as the amortization period of the project on account of the economies made in the power price cost block amounted to only a little more than a year. Noteworthy restrictions of the firm's production process were not observed.

Nevertheless it should be emphasized that the form of switch-off strategy described here, with reference to a case study, is simplistic in as far as only the past consumption development in connection with the fixed limit values is brought into play for the switching on and off of equipment. A prognostication is not, however, made of the remaining consumption development within the respective 15 minutes, taking the prevailing production situation into account.

Thus in addition to the switch-offs which are necessary in the case of correct regulation and the resulting disruption possibilities in the production process, the method also involves a high risk of wrongly timed interventions: If equipment is turned off belatedly the disruption to the production process is increased because emergency switch-offs are subsequently necessary. It is also not possible from the outset to exclude the possibility in every case that the maximum prescribed average power consumption will be exceeded. On the other hand switch-offs can occur that in retrospect prove to be premature or not necessary at all. Finally there is also the danger that the switch-ons can be subject to incorrect control measures of a similar nature.

(b) Switch-Offs with Explicit Consumption Prognostications

It is to be expected that the prevailing state of the production process can be better adapted to when explicit consumption prognostications are employed. A method of this sort has been developed for the electricity consumption of a West German iron and steel plant (VOSS and WARTMANN 1977). The incentive behind smoothing the electricity utilized is to be found here in the agreement with the energy supply company which fixed an upper limit for the quarter hourly mean of the power consumed. If this limit is exceeded this results in considerable extra costs.

This control concept provides that the quarter hourly mean is predicted – in regular intervals of, for example, one minute – on the basis of the prevailing actual consumption level. Thus if the consumption is in danger of exceeding the power consumption limit certain sets of machines or parts of the firm can be temporarily switched off in accordance with a predetermined priority sequence. In this case equipment is not switched on again within the same 15 minute interval in which it has been switched off. The prognostications are carried out either with help of the method of exponential smoothing, which is relatively simple to implement, or with the help of consumption patterns, identified as being characteristic in the individual parts of the plant. The latter method is far more difficult to execute. The thus implemented regulation of the purchase of electricity is clarified in Fig. 6.3.

In every measurement period, on the basis of the actual consumption which becomes successively known for the quarter hourly average power consumption, a series of estimated values is ascertained at the end of which the actual 15 minute mean is revealed. There can be a considerable discrepancy between this actual value and the first estimated value of the subsequent measurement period which is being simultaneously calculated but for which actual consumption information is not yet available.

Within the company an internal switch-off limit has been fixed below the limit arranged with the energy supply company which should not be exceeded by the quarter hourly average and which thus serves as the determining factor in the triggering off of the regulation measures. The switch-off signal is not, however,

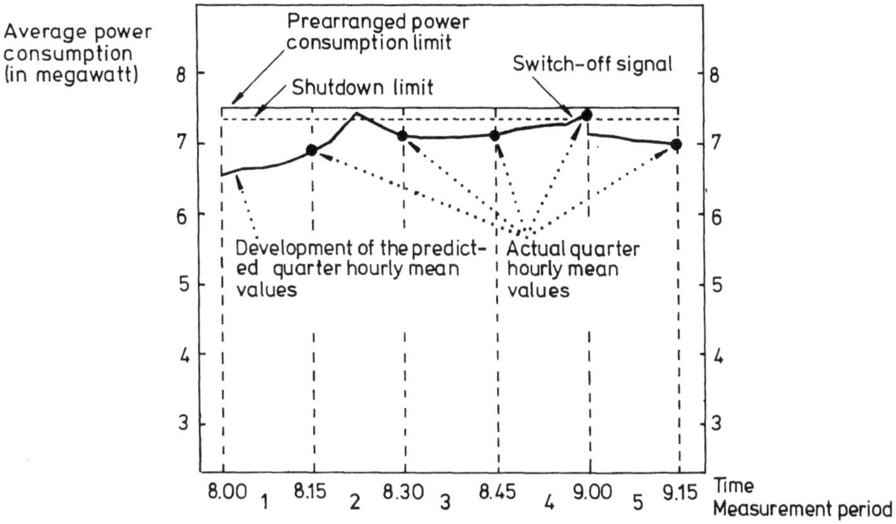

Fig. 6.3. Estimated and actual quarter hourly mean of the power consumption in the case of regulating the purchase of electricity by means of a switch-off strategy with explicit prognostications

of necessity given every time that the estimated values exceed this internal power consumption limit: If it turns out when this mark is exceeded that an intervention at a later point in time would presumably still be in good time then for the time being nothing is undertaken. It is possible that at a later point the estimated average value sinks to a lower level below the switch-off limit without intervention so that a premature regulation measure could be avoided. A case of this sort is shown in Fig. 6.3 in the second measurement period.

On the other hand, however, it can also happen that the internally fixed power consumption limit is exceeded although the switch-off signal has been effected. The primary explanation for such an occurence is that when the switch-off signal is given for a consumer the operation which is in process or which is just about to begin is first carried out so that the desired relief effect does not come into effect for a certain period of time. If then, for example, the electricity consumption unexpectedly increases in the last minutes of a quarter of an hour the switch-offs for this period of time can prove to be belated – but for the next 15 minute interval, as a rule, they cause the estimated values to fall below the switch-off boundary. A situation of this sort is assumed in the fourth and fifth measurement periods in Fig. 6.3.

In spite of the methodical refinement in comparison with the switch-off strategy without explicit prognostications the regulation form presented here also contains weaknesses. Thus the switch-offs carried out normally result in production disruptions of varying severity, a consequence that is intensified all the more by the fact that the risk of intervening unnecessarily or at the wrong time cannot be fully ruled out. In particular belated adjustment measures undertaken to reduce the quarter hourly mean involve the risk of emergency switch-offs or the danger that the power consumption limit arranged with the supply company will be exceeded. Apart from that the implementation costs of this method, depending on the selected method of prediction can be very high, above all because of the requirements which have to be met when equipping the necessary computing system with hardware and software.

III. Control of the Purchase of Pipe- and Line-Based Fuels with the Help of Machine Schedules

The control of the purchase of pipe- and line-based fuels with the help of consumption stoppages can indeed lead to a detectable fall in production costs. It is, however, not without its weaknesses as has been demonstrated particularly in connection with the switch-off strategy where there is a limited capacity for storing the energy source. For this reason, taking electricity as the example, another concept of smoothing energy consumption is to be examined. This examination concerns machine schedules by means of which the time at which the production equipment is switched on is preset so that switch-offs which cannot be predetermined and which if necessary have to be carried out at short notice can be eliminated.

The first step in this direction is the prevailing modus operandi in company practice which, without employing a special decision orientated quantitative regulation method, "manually" draws up a "timetable" of suitable equipment for the relief of the daily peak consumption period, i.e. the working times of the selected sets of machines are, as far as possible, laid outside the expected phase of the maximum power consumption. The application of this procedure presupposes that the course of a day's typical consumption development is known. This information can be ascertained on the basis of repeated measurements of the daily profiles of power consumption (Wolobrinski and Sorin 1976).

Subsequently a new method is to be presented and substantiated with exemplary calculations with an aim to developing the procedure which has just been outlined. In particular an analytical procedure – the mixed integer linear programming – should be applied as a methodical back-up.

1. Starting Situation and Characteristics of a Model for the Machine Scheduling

First of all it has to be clarified to which starting situation the regulation model is being applied and how it is built up in terms of its characteristic features:

The smoothing of the production plant's consumption of electricity should interest the company in as far as the highest average power consumption values ascertained during the measurement period are partly decisive for the size of the total electricity costs – a measurement period covers a quarter of an hour as is usual in accounting practice. It is unnecessary to make assumptions about further details of the price settlement, for example as to whether a power consumption limit is to be arranged with the energy supply company or whether a choice should be made between a flat and a steep price settlement with a different relationship between the work price and the power consumption price (for information about the various common electricity price settlements see FINSINGER and KLEINDORFER 1981; KILGER 1981, p. 386; VERBAND DER ENERGIEABNEHMER 1978, p. 3). The decision making process with regard to this aspects can possibly be supported by the results of the model analysis to be presented in the following.

The production planning is predetermined to a large degree, i.e. the production programme, the production potential, and the production procedure in this framework cannot be influenced in any way, and the process planning, apart from the utilization times of the electricity consumers which have been selected for regulation purposes, is also stipulated. Upon this basis, in particular also for given intensities, the power demand of the individual machines in the production process can be indicated as a quarter hourly average.

The machines are differentiated according to whether they have to be switched on at a certain time in the course of the day or not. Accordingly the whole electricity demand is then divided up into the two components of autonomous and controllable consumption.

The average quarter hourly power consumption is planned for the course of a representative day in order to smooth the electricity demand as much as possible: First of all the development of the autonomous electricity consumption is to be established on the basis of the existing production plan. Then the structure of the controllable electricity consumption is laid down in such a way that the maximum power consumption in the course of the day is minimized. This affects those machines for which there is a scope for decision with regard to their operating phases, but not for the total operating time. Technical and economic production requirements which can be expressed in the ban on the simultaneous operation of otherwise controllable machines will as a rule restrict the freedom to regulate the operating times.

Thus this regulation procedure which is illustrated in Fig. 6.4 is organized so as to avoid stoppages with a view to smoothing the purchase of electricity

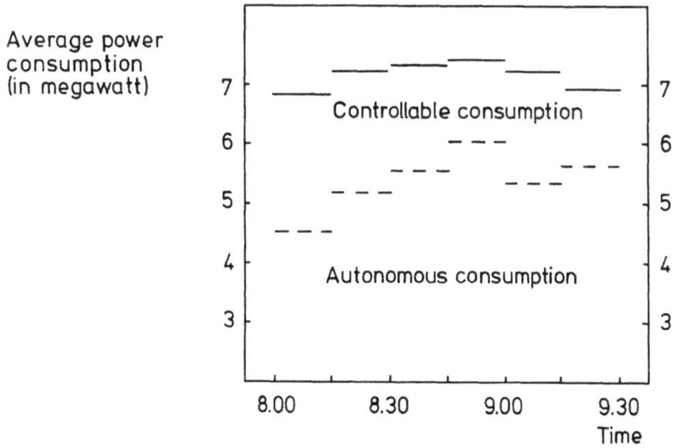

Fig. 6.4. Planned quarter hourly mean values of the power consumption where the purchase of electricity is regulated by means of a machine utilization programme

from the outset. It can, however, also be used in connection with a switch-off strategy or, more broadly speaking, in connection with a control method with purchase stoppages, in which case it is brought into play in order to usefully lay down intervention limits.

2. Formulation of the Model

For the formal construction of the model to regulate the purchase of electricity with the help of a machine utilization programme let be denoted by:

h or $h' = 1, \ldots, H$ electricity consumers/machines,

$t = 1, \ldots, T$ measurement periods (each one lasting a quarter of an hour),

v_{th} operating variable of the machine h for period t ($v_{th} \in \{0, 1\}$),

$e_{\max}$ purchase spike variable as the highest power consumption which occurs on average in a measurement period as a result of the regulation,

e_t^{aut} autonomous power consumption on average of the measurement period t as sum of the consumption of all machines which have to be switched on in t,

e_h^{disp} electricity consumption on average of the machine h whose operating time is at least controllable within certain limits,

g_h the total required operating duration of the controllable machine h expressed in measurement periods,

δ parameter for the characterization of the length of an interval of time in measurement periods,

η parameter for the characterization of the minimum operating duration in measurement periods,

θ auxiliary index for the characterization of periods,

τ_h period of the latest operating start of the machine h,

Ψ_ν sub-quantity from the totality of all machines or machine pairs ($\nu = 1, \ldots, 9$; $\Psi_1, \ldots, \Psi_5 \subset \{1, \ldots, H\}$; $\Psi_6, \ldots, \Psi_9 \subset \{1, \ldots, H\} \times \{1, \ldots, H\}$),

Ω_{ν_h} or $\Omega_{\nu_{(h,h')}}$ sub-quantity from the totality of all measurement periods, dependent on the machine h or the machine pair (h, h') from the quantity Ψ_ν ($\nu = 2, 3, 6, \ldots, 9$; $\Omega_{2_h}, \Omega_{3_h}, \Omega_{6_{(h,h')}}, \ldots, \Omega_{9_{(h,h')}} \subset \{1, \ldots, T\}$).

With these symbols the following model can now be formulated. It comprises a basic framework which is necessary at all events – objective function (0) and restrictions (1) – (3) – and a series of additional restrictions (4) – (14), which, depending on the facts of a concrete problem situation, are possibly to be taken into account:

$$
\begin{array}{llll}
(0) & e_{\max} & \rightarrow \text{Min!} & \\
(1) & e_{\max} & \geq e_t^{\text{aut}} + \sum_{h=1}^{H} v_{th} e_h^{\text{disp}}, & t = 1, \ldots, T , \\
(2) & \sum_{t=1}^{T} v_{th} & = g_h , & h = 1, \ldots, H , \\
(3) & v_{th} & \in \{0,1\} , & h = 1, \ldots, H , \quad t = 1, \ldots, T , \\
\hline
(4) & \sum_{t=1}^{\tau_h} v_{th} & \geq 1 & \text{for all } h \in \Psi_1 , \\
(5) & v_{th} & = 0 & \text{for all } h \in \Psi_2 , \ \text{for all } t \in \Omega_{2_h} , \\
(6) & v_{th} & = 1 & \text{for all } h \in \Psi_3 , \ \text{for all } t \in \Omega_{3_h} , \\
(7) & \sum_{t=\theta+1}^{\theta+\delta} v_{th} & \geq \eta & \text{for all } h \in \Psi_4 , \ \theta = 0, \delta, 2\delta, 3\delta, \ldots, T - \delta - (T) \bmod \delta , \\
(8) & v_{2h} & \geq v_{1h} & \text{for all } h \in \Psi_5 , \\
(9) & v_{t-1,h} + v_{t+1,h} & \geq v_{th} & \text{for all } h \in \Psi_5 \ \ t = 2, \ldots, T-1 , \\
(10) & v_{T-1,h} & \geq v_{Th} & \text{for all } h \in \Psi_5 , \\
(11) & v_{th} + v_{th'} & \leq 1 & \text{for all } (h,h') \in \Psi_6 , \ h \neq h' , \ \text{for all } t \in \Omega_{6_{(h,h')}} , \\
(12) & v_{th} + v_{th'} & \geq 1 & \text{for all } (h,h') \in \Psi_7 , \ h \neq h' , \ \text{for all } t \in \Omega_{7_{(h,h')}} , \\
(13) & v_{th} - v_{th'} & = 0 & \text{for all } (h,h') \in \Psi_8 , \ h \neq h' , \ \text{for all } t \in \Omega_{8_{(h,h')}} , \\
(14) & v_{th} + v_{th'} & = 1 & \text{for all } (h,h') \in \Psi_9 , \ h \neq h' , \ \text{for all } t \in \Omega_{9_{(h,h')}} .
\end{array}
$$

The case in hand constitutes a minimax problem in that the maximum power consumption which occurs on average in a measurement period should be minimized; the objective function (0) gives expression to this minimization instruction with regard to the purchase spike. According to (1), on the other hand, the overall demand per quarter of an hour, which is made up of autonomous and controllable components, forms the lower limit of $e_{\max}$, which is to be minimized; the variables v_{th} each indicate whether the machine h should be switched on in t – $v_{th} = 1$ – or switched off – $v_{th} = 0$. By interacting (0) and (1) $e_{\max}$ is then attributed the exact value of the highest average demand and the controllable consumption is regulated in such a way that this maximum purchase turns out to be as low as possible.

On account of (2) it is guaranteed that the total of the scheduled operating times of the machines brought in for regulation corresponds with the overall required operating duration. The restrictions (3) contain the binary definitions of the operating variables.

The basic framework of the model has already been set up with the objective function (0) and the restrictions (1) – (3). The additional restrictions, (4) – (14), can make allowance for a number of other technical and economic requirements which might have to be observed when setting up the machine utilization programme.

The restrictions (4) – (10) can be formulated for some machines h but in an extreme case all or none of the machines are affected; with regard to the machine pairs this fact is equally valid for the restrictions (11) – (14). The restrictions (5), (6) and (11) – (14) in the case of every machine h or machine pair (h, h') have to be formulated at least for one period t.

The framework of the utilization time planning can stipulate, for example, the latest time at which a machine to be put in operation, which can be ensured by (4). Other restrictions cover the operating periods and the standstill periods which are prescribed from the outset for equipment which is otherwise not fixed in terms of time – this can be expressed by (5) or (6) respectively. Thus the latest possible processing deadline requested for a controllable machine which is still within the planning period is accommodated for with the help of (5) by setting the relevant operating variables of the subsequent measurement periods at zero.

(7) describes the requirement that sets of machines in consecutive intervals of equal length, the duration of which is indicated by δ, must be switched on for at least η measurement periods. Thus, for example, where $\delta = 4$ and $\eta = 1$ within every hour they must be in operation for at least a quarter of an hour. Furthermore if it is assumed that the planning period comprises $T = 34$ measurement periods then the auxiliary index θ, which here notates the measurement period before the hour which is just about to be observed, takes the values 0, 4, 8, 12, ..., 28. The value 28 is calculated in accordance with (7) by means of $T - \delta - (T) \bmod \delta = 34 - 4 - 2$; $(T) \bmod \delta$ indicates the remainder of measurement periods for which the restrictions no longer have to be fulfilled.

Frequently it will be necessary or desirable to observe a minimum operating time without interruption for a set of machines. A minimum duration of two measurement periods, i.e. half an hour, is expressed in the restrictions (8) – (10). (8) and (10) ensure that an isolated switch-on can not be scheduled for the first or for the last measurement period, whilst the other 15 minute intervals are included in (9). If a minimum operational phase of more than half an hour is considered the number of restrictions is increased but without causing any fundamental difficulties.

In the remaining restrictions, which are formulated here by way of an example for two machines, formal expression is given to the requirement that forbids simultaneous operation or standstill – (11) and (12) respectively – and to the requirement that demands the same or different operating phases – (13) and (14) respectively. It is quite possible to formulate these conditions for more than two machines.

In order to fulfil the conditions of a concrete case it is possible, particularly in the restriction part (4) – (14) of the model, that modifications or additions, for example the integration of longer minimum operating times without interruption, prove to be useful or necessary. In the exemplary calculations in the following section reference will be made to the restrictions of the model version which has been set up here.

3. Notes on the Model Solution and Exemplary Calculations

The arithmetical execution of the model is not unproblematical in as far as, with one exception – namely e_{max} –, only 0-1-variables have to be processed, i.e. the problem at hand is essentially one of binary linear programming. As it to be shown in detail in the following expositions, and particularly in the second exemplary calculation, it is possible, by using the corresponding standard software, to master a problem of practical relevance which is characterized by a relatively large number of 0-1-variables. It is, however, quite conceivable that there are cases in which a calculation time, which can be regarded as tolerable, will probably be exceeded.

The structure of the model does, on the other hand, contain possibilities of simplifying the calculation process in the form of variable groups within which the number of the acceptable value combinations is considerably reduced in comparison with the number of basically conceivable combinations. In the literature the most radical expressions of such sub-quantities of variables, which are characterized by special limitations, are commonly defined as special ordered sets. Thus a group of variables is defined as special ordered set of type 1 within which precisely one variable must assume a value other than zero. This type 1 generally occurs among 0-1-variables (WILLIAMS 1978, p. 164).

Even if one now merely looks more closely at the indispensible basic framework of the model which has been formulated here, i.e. (0) – (3), where the restrictions

$$(2) \qquad \sum_{t=1}^{T} v_{th} = g_h \, , \qquad h = 1, \ldots, H \, ,$$

are imposed, then one comes across a very similar, but somewhat less strict form of the grouping. Per machine h the number of operating plans, which fulfil condition (2), amounts to $\binom{T}{g_h}$; without this restriction there would be 2^T possibilities. Where $T = 10$ and $g_h = 3$, for example, the number of value combinations which come into question for this g_h are narrowed down to 120 from 1024 basically conceivable combinations. Where $T = 10$ and $g_h = 2$ the reduction of solutions from 1024 to 45 turns out to be even greater, and finally where $g_h = 1$ a special ordered set of type 1 is given with 10 remaining acceptable solutions.

In the additional part of the model similar restrictions, which greatly reduce the number of feasible solution points, can occur. If such structure components are used by means of a corresponding form of algorithms, which are incorporated in the software – which can be expected for the future –, then far more complex cases than those examined in the following can be mastered with an acceptable level of effort (see also CROWDER et al. 1983).

The first exemplary calculation to be presented is to be based on a problem which, for the sake of comprehensibility, merely comprises ten measurement periods, i.e. two and a half hours. With the help of this problem it is simple to show that successive solution methods for the derivation of machine utilization programmes often lead to sub-optimum schedules. Subsequently a more comprehensive case will then be dealt with.

For the first example let the autonomous electricity consumption of the production equipment from the first to the tenth quarter hourly interval be on average 6.9, 6.5, 6.4, 6.75, 6.55, 7.0, 6.5, 5.8, 5.0, and 6.6 megawatts. Four machines which can be controlled in terms of time are available whose power consumption is on average 1.5 (machine 1), 1.0 (machine 2), 0.7 (machine 3), and 0.5 (machine 4) megawatts and which have total operating times of 3 (machine 1), 5 (machine 2), 6 (machine 3), and 7 (machine 4) measurement periods.

Apart from that the following restrictions should be included in the formulation. The processing steps on machine 1 have to be completed in the seventh sub-period at the latest. Machine 2 must be switched on at least once every half an hour, beginning with the 30 minute interval 8.00 – 8.30. Machine 3 has to be in operation for at least two sub-periods without interruption at any one time. The machines 1 and 3 must not run together. During each of the first six sub-periods machines 2 and 4 must exhibit the same operating mode at any one time.

If the machine utilization occurs without a smoothing of the purchase of electricity being carried out, and if the machines 1, 2, and 4 are, therefore, switched on in the first measurement period then this results in an average power consumption of 9.9 megawatt in this first interval. In fact it is frequently the case in industrial practice that at the beginning of a working day or shift an above average electricity consumption or even the spike will occur (see also VERBAND DER ENERGIEABNEHMER 1978, p. 6).

In comparison with the machine utilization without smoothing the purchase of electricity, certain successes can certainly frequently be achieved by means of a regulation with the help of a successive planning procedure. This method should also be applied here first of all in the following way. The controllable machines are allocated one after another in order of decreasing electricity intensity, i.e. beginning with the largest electricity consumer, to the measurement periods with the lowest consumption until the required total running times have been reached. The rest of the restrictions of the machine utilization programme will also be gradually included, namely upon arrival at the solution step at which it is unavoidable.

In the case in hand, when arranging the operating sequences attention must, therefore, firstly be paid to the fact, that machine 1 must not run in the last three measurement periods – the fact that it is forbidden for machine 1 to be in operation at the same time as machine 3 will, on the other hand, not be considered until the utilization of the latter is planned. It must be provided for that machine 2, which is then included, is switched on at least once every half an hour, in which case the consumption of machine 1, which has been fixed in terms of

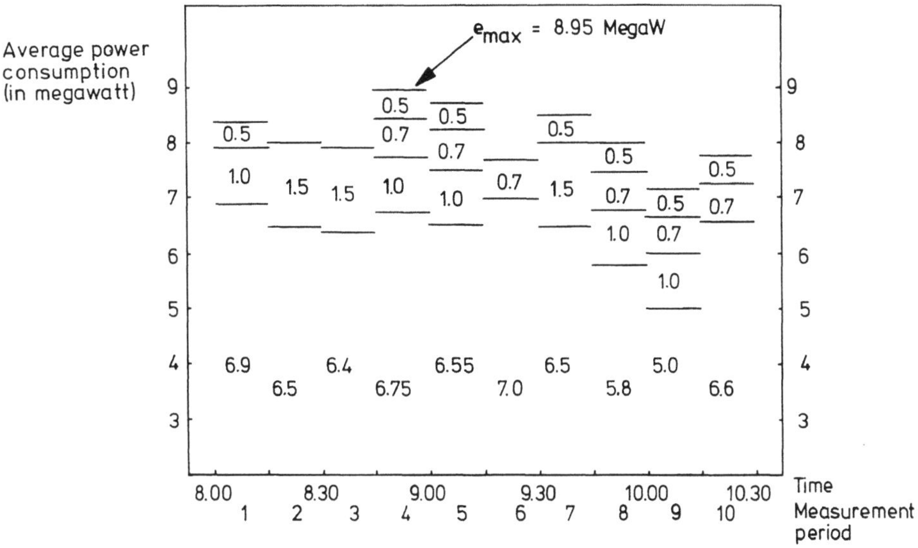

Fig. 6.5. An acceptable machine utilization programme, ascertained by means of a successive solution method, for a problem with ten measurement periods and four machines for regulation

time, is now to be regarded as data. Thus the operating times of all four controllable machines are progressively laid down.

This successive 'filling up' results in an acceptable solution with a maximum consumption of 8.95 megawatt as is shown in Fig. 6.5.

The model formulated in the preceding section was employed for the simultaneous planning with the help of which the optimal solution can be reliably found. This yielded 40 binary variables and 45 restrictions – if the binary definitions of the operating variables are left out of account – (for a complete model formulation see WOLF 1987, p. 238). At this point an overview should be presented of the restrictions of the model by means of which the special assumptions and limitations of the first exemplary case were recorded. Table 6.1 serves this purpose.

The model calculation was carried out on a CDC 175-system with the help of the standard software system APEX-III and required a CPU time of less than 3 seconds. The optimal solution exhibited a consumption peak of only 8.7 megawatt, as can be seen in Fig. 6.6.

The final deciding factor for the reduction of the electricity purchase costs, which can be achieved by means of consumption control, is the contractual agreement made with the energy supply company. If, for example, a constant power consumption price of DM 15.– per kilowatt and per month is assumed this results in a difference in monthly costs between the machine utilization without smoothing (maximum power consumption of 9.9 megawatt) and the optimal solution (maximum power consumption of 8.7 megawatt) of DM 18,000.–. Compared with the 'timetable' in the case of successively introduced smoothing (8.95 megawatt maximum power consumption) the optimal solution results in a cost difference of DM 3,750.–.

Table 6.1. Overview concerning the application of the approach to the conditions of a problem with ten measurement periods and four machines which are brought in for regulation

Assumptions/restrictions of the first exemplary case	Restrictions of the approach
Autonomous power consumption of the then measurement periods, power consumption of the four controllable machines	(1)
Total operating time of the four controllable machines	(2)
Binary definitions of the operating variables	(3)
Latest production deadline for machine 1	(5)
Minimum operating time per half an hour for machine 2	(7)
Minimum operating time without interruption for machine 3	(8) – (10)
Ban on the simultaneous operation of machines 1 and 3	(11)
Requirement that during the first six measurement periods machines 2 and 4 are at the same operational stage	(13)

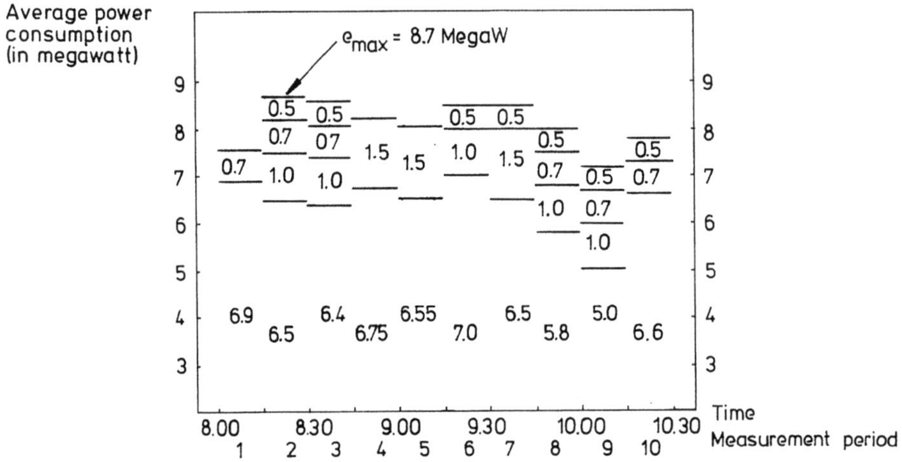

Fig. 6.6. The optimal machine utilization programme, ascertained by means of the simultaneous solution procedure, for a problem with ten measurement periods and four machines for regulation

The following deals with a second exemplary case based on a more comprehensive problem, which covers the total duration of a one-shift production. In this case the operation phases of three controllable machines are to be determined for thirty-four measurement periods.

The autonomous electricity consumption from the first to the thirty-fourth quarter hourly interval amounts to 8.5, 9.0, 9.1, 9.0, 8.8, 8.6, 8.8, 7.9, 8.9, 7.9, 7.7, 8.1, 8.5, 8.0, 7.5, 7.3, 7.6, 7.0, 7.5, 7.6, 7.2, 7.5, 8.0, 7.6, 7.9, 8.2, 8.2, 8.3, 8.5, 8.7, 8.5, 8.2, 8.4, and 8.4 megawatts. The three controllable machines exhibit a power demand of 1.6 (machine 1), 0.9 (machine 2), and 0.6 (machine 3) megawatts. The required total operating times are 18 (machine 1), 22 (machine 2), and 20 (machine 3) measurement periods.

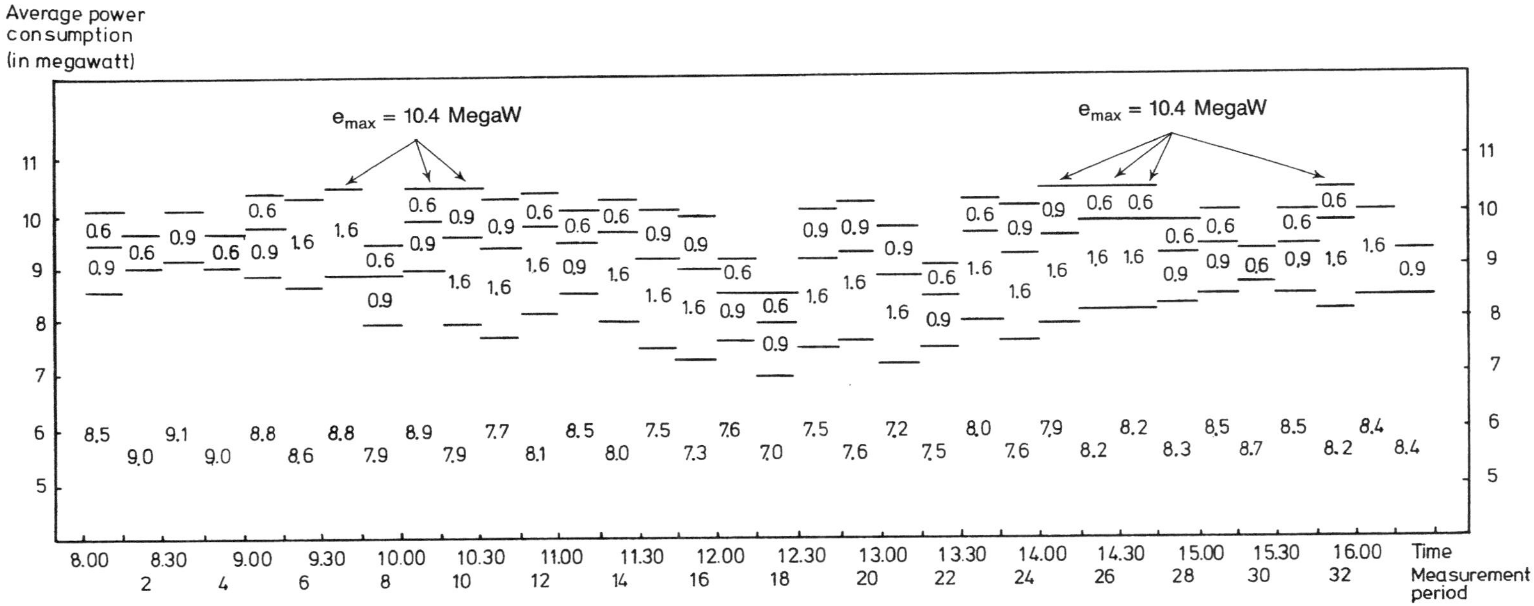

Fig. 6.7. A machine utilization programme, ascertained by means of a simultaneous procedure, for a problem with thirty four measurement periods and three machines for regulation

In addition the following restrictions on the utilization planning are assumed: machine 1 must run each time for at least half an hour without interruption. In the case of machine 2 it must be switched on at least once every half an hour, beginning with the 30 minute interval from 8.00 – 8.30. Machine 3 must be in operation for at least half an hour at any one time without a break and apart from that every full hour, from 8.00 till 16.00, it must be in operation at least twice. Finally machines 1 and 3 are not allowed to run together in the time from 11.30 – 13.30.

In the solving of this problem the approach set up in the preceding section is applied anew. This resulted in a model with 102 binary variables and 138 restrictions – without the binary definitions of the operating variables (for the complete model formulation see WOLF 1987, p. 241).

The calculation was executed on the CDC 175-system, which was also used for the first example, and again with the help of the programme system APEX – this time in the APEX-IV version. So as not to prolong the required CPU-time too much the calculation process was broken off after 13 minutes. The optimality proof could thus not be furnished for the results which had been found after about 11 minutes. The quality of the solution does, however, appear to be very high, particularly since the applied branch-and-bound-procedure produced the statement that the objective function value could be improved at the most by 3.1%.

Figure 6.7 shows the corresponding machine utilization programme. It is made clear here that the consumption peak of 10.4 megawatt occurs seven times.

The application of the successive solution method described in connection with the first exemplary case, on the other hand, leads to a consumption peak of 11.0 megawatt. If in turn a constant power consumption price of DM 15.– per kilowatt and per month is assumed then the monthly cost difference between this solution and that found with simultaneous procedure amounts to DM 9,000.–.

Thus the setting up of the machine utilization programme for smoothing the purchase of electricity can lead to noteworthy reductions in costs. The simultaneous procedure presented here is more likely to be superior to a successive solution process the more the freedom in planning the utilization time is restricted by technical and economic limitations. It is, however, quite possible to use this method in connection with the other pipe- and line-based fuels, namely gas and district heat. Thus, for example, it is conceivable that the planning of quantities of purchased gas could be carried out in accordance with the procedure presented here. Over and above that in the case of this source of energy it would be basically possible to avoid unexpectedly exceeding the planned purchase maximum without causing production disruptions by, when necessary, undertaking purchase stoppages and at the same time using storage capacities.

References

Crowder, H., et al.: Solving Large-Scale Zero-One Linear Programming Problems, in: Operations Research, 1983, pp. 803 – 834.

Fatti, L.P.: Optimal Smoothing of Demand for Industrial Gas, in: Journal of the Operational Research Society, 1983, pp. 583 – 590.

Fatti, L.P.: Modelling a Reservoir with Stochastic Outflows – Application to the Optimal Smoothing of Demand for Industrial Gas, in: Operational Research, 1984, pp. 864 – 873.

Finsinger, J., and Kleindorfer, P.R.: Höchstlast- und Spitzenlastpreisbildung. Ein Vergleich der Strompreise für typische Sonderabnehmer in der Bundesrepublik Deutschland und Frankreich, in: Zeitschrift für Energiewirtschaft, 1981, pp. 48 – 55.

Hugel, G., and Schmitz, H.: Betriebliche Energiewirtschaft, Berlin-Köln 1977.

Kern, W.: Aktuelle Anforderungen an die industriebetriebliche Energiewirtschaft, in: Die Betriebswirtschaft, 1981, pp. 3 – 22.

Kern, W.: Konzepte zur Energiebewirtschaftung in industriellen Betrieben, in: Betriebswirtschaftliche Forschung und Praxis, 1984, pp. 107 – 119.

Kilger, W.: Flexible Plankostenrechnung und Deckungsbeitragsrechnung, 8th edition, Wiesbaden 1981.

Richarts, F.: Energieeinsparung und verbesserte Energieausnutzung im industriellen und gewerblichen Bereich, in: IHK zu Köln (ed.): Energie wirtschaftlicher einsetzen, Köln 1976, pp. 73 – 92.

Verband der Energieabnehmer (ed.): Energietips für Einkauf und Betrieb, written by G. Bischoff et al., Hannover 1978.

Voss, G., and Wartmann, R.: Untersuchung von Abschaltstrategien zum Vermeiden von Stromverbrauchsspitzen, in: ZfbF-Kontaktstudium, 1977, pp. 83 – 90.

Williams, H.P.: Model Building in Mathematical Programming, Chichester et al. 1978.

Wolf, J.: Industriebetriebliche Energienutzung und Produktionsplanung, Heidelberg-Wien 1987.

Wolobrinski, S.D., and Sorin, B.P.: Die Regelung der täglichen Belastungskurven von Industrie- und Verkehrsbetrieben, in: Energietechnik, 1976, pp. 459 – 461.

Part III

Joint Production with Surplus, Waste and Hazardous Byproducts

Chapter 7

Optimal Heat-Matched Cogeneration of Energy in a Firm Owned Power Station A Case Study

GÜNTER FANDEL and JOACHIM REESE

I. Introduction

Energy represents for every industrial site a fundamental process input which can only be substituted with great difficulty and without which industrial manufacture in general is impossible. The requirement is hereby concentrated on the utilization of distinct types of thermal energy for the process steam and heating, as well as for power and light. According to WOLF (1987) the total demand in 1981 for energy for industrial purposes in the Federal Republic of Germany was divided as follows – 75.9% for the manufacturing process, 10.9% for space heating, and 13.2% for power and light. Although serveral energy resources were used equally for the production of heat, power and light were generated almost without exception by means of electricity. Heat as well as power and light, or rather electricity, are all commodities which have only a limited storage capacity. This leads to considerable procurement problems when the energy requirement fluctuates.

In the case of the external supply of all types and amounts of end-use energy the difficulties often lie in peak demands, because they lead to considerable extra cost. Even if the peak values are only sporadic, i.e. they occur for relatively short periods of time, they nevertheless make a substantial contribution to the overall cost in the firm in that they necessitate standby quantities at the supplier which cause in turn an according standby tariff. This is why the main concern must be to subject the period requirements to regular checks with a suitable operational policy in mind. According to this suitable instruments are either temporary delivery cuts or special energy usage programs. In spite of the many proposals in this area (e.g. CONSTANTOPOULOS et al. 1983; FATTI 1983 and WOLF 1987) the results are often usually unsatisfactory because breaks in production occur or unforeseeable demand cannot be calculated with sufficient accuracy.

For this reason it is often recommended that the problem basis should be broadened to include the self generation of energy. The decisions to be taken here are of a completely different nature and can basically be divided into three areas:

- investment decisions, i.e. which facilities for self generation should actually be provided in the long run. This fundamental criterion, directly concerning the energy policy of the enterprise, has led to various proposals regarding the assessment of the necessary units (e.g. SKEA 1981; PORTER and MASTANAIAH 1982, and many others). These offer static and dynamic methods of invest-

ment calculation and take the special requirements of the whole energy problem into account.
- process option decisions with respect to the generating technology, i.e. under which conditions which units should be employed to produce energy.
If a decision has been taken in favour of a self generation of energy and if the units have already been procured, the question still remains as to the economic efficiency of the individual units regarding periodic demand and other basic conditions, given of course that there are more than one boiler, turbine etc. with varying capacities and efficiencies. This subject has special relevance where multi-stage production processes can be observed, i.e. also in power stations. It is a general topic of discussion in the relevant literature, but in the area of energy there are only isolated references to the specific problems involved (e.g. STOCKS et al. 1985).
- market decisions, i.e. the economics of additional transactions on the market alongside self generation of energy.
This area is important not only as regards procurement but also as regards sales. It includes considerations as to whether at least part of the energy demand should be supplied from outside, or whether surplus production should be fed into the public network. According to their nature the respective decisions can be made in the short run or in the long run. At the same time they influence and increase the problem areas of investment and process options in that more alternatives become worth testing because of the relaxation in the conditions of supply. In the relevant literature market decisions are usually, logically, treated as a part of the respective fundamental problem of the procurement of a facility or of its utilization.

Long-term investment decisions are not of importance in the economic analysis carried out here, because the units necessary for self generation of energy are present in the power station and do not have to be replaced. It is more important to view the employment of the different units dependent on the periodic demands. If the purchase of energy is to be regarded hereby as an admissible process alternative in a wider sense, then the economic analysis will relate to process decisions as well as to market decisions insofar as the procurement aspect is concerned.

The efficiency of a power station attached to a plant and used to supply energy for production in a medium-sized firm in the electrotechnical industry is to be tested with the aid of the classical stage-oriented calculation (see e.g. ALBACH 1962). As the power station is integrated within the organization of the firm as a cost center, within which the internal energy requirements are to be guaranteed at the most advantageous costs possible, and does not have to produce any market performance of its own, the analysis has accordingly to limit itself to an investigation of the costs incurred. Certainly, the outside supply of electricity will be included in the study, but the possibility of releasing energy into the public network is left out of the considerations, as is the harmonizing of the firm's production program with the economic generation of energy. The data used in the study were collected over the period of a calendar year. Because of the extreme seasonal and shift-based fluctuations in energy demands the data were at first arranged according to suitable periods. Finally, special calculations for each period of the year observed were carried out before the focus was centered upon the procedural policy for the whole year.

II. Description of the Problem

The power station produces as final products steam of 10 bar and 1.5 bar for the manufacturing process and for heating, as well as electricity for use in various areas within the plant. Complete self generation of the required end-use energies is not possible as the production of heat and electricity are coupled in such a way as to make an exact coordination of the production levels with the requirements data impossible. In practice, this problem of cogeneration of thermal and electric energy is taken into account by first of all using the units to produce heat. After the given volume of process and space heat has been supplied, the amount of electricity is then given automatically by the joint production and has to be increased through outside supply in order to meet the respective current demand. It is possible that the self generation of electricity in larger amounts would be more profitable, but this would cause pass-out steam, which would have to be disposed of. In view of the resulting costs of the disposal, and of the fact that equipment for recycling the steam as condensed water is not at present available, it is not necessary to discuss such a policy any further.

The analysis is rather more concerned with short-term objectives, and is aimed at the operational policy of the power station on an hourly basis. All cost estimates are based on peak demand values which are regarded by the management as representative of the respective season and shift (Table 7.1). The fluctuations in demand within the periodic delimitations are so small that they can be

Table 7.1. Peak demand (per hour)

	November – March	July – August	April – June/ September – October
Early shift (6 – 14)	17 t 24 t 8.6 MegaW (Period I)	10 t 6 t 7.7 MegaW (Period IV)	13 t 15 t 8.3 MegaW (Period VII)
Late shift (14 – 22)	14 t 21 t 8.6 MegaW (Period II)	8 t 5 t 7.7 MegaW (Period V)	10 t 14 t 8.3 MegaW (Period VIII)
Night shift (22 – 6)	14 t 20 t 6.5 MegaW (Period III)	7 t 5 t 6.9 MegaW (Period VI)	10 t 14 t 7.2 MegaW (Period IX)

Top lines for each shift: 10 bar steam; middle lines: 1.5 bar steam; bottom lines: electricity.

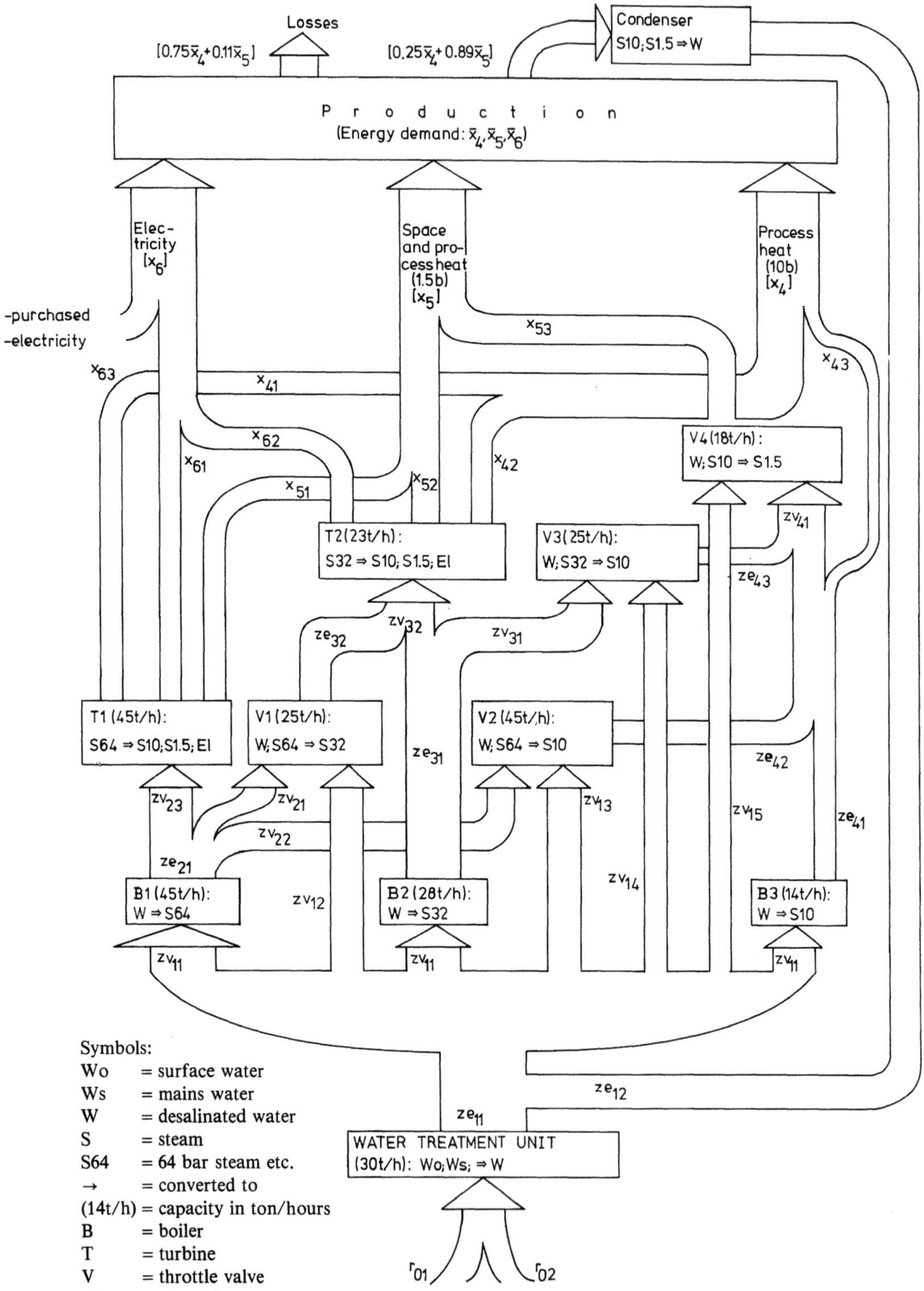

Symbols:

Wo	= surface water
Ws	= mains water
W	= desalinated water
S	= steam
S64	= 64 bar steam etc.
→	= converted to
(14t/h)	= capacity in ton/hours
B	= boiler
T	= turbine
V	= throttle valve
El	= electricity

Fig. 7.1. Power station flow chart

ignored, so the deviations between the peak values and the average values are no longer relevant for the purposes of the analysis. As can be seen from Table 7.1, the shifts are grouped in rows and the seasons in columns. The nine periods of time which can thus be deduced are numbered consecutively and will be judged separately.

The generation of the end products process heat (10 bar and 1.5 bar), space heat (1.5 bar), and electricity takes place by means of a multi-stage production process which can be exclusively described by means of linear technologies, and which ends in a joint production of all end products at the final stage.

At the first level, completely desalinated water is produced from surface water, mains water and recycled condensate. This can be converted into steam at the next level by means of three oil-fired boilers, B1, B2, and B3 alternatively. B1 generates steam at 64 bar, B2 at 32 bar, and B3 at 10 bar. The steam pressure is then possibly throttled before the steam can be consumed in the generation of electricity or directly used in the production area of the firm. The throttle valves V1, V2, V3, and V4 are available for pressure reduction; the generation of electricity can take place using two turbines, T1 or T2, which have different capacities and different efficiencies. The utilization of the turbines causes a cogeneration of electricity and steam, i.e. the amount of electricity produced results at the same time in the generation of steam at 10 bar and 1.5 bar in fixed proportions.

The whole structure of the present problem is shown graphically in Fig. 7.1. The energy flows are identified by means of symbols which are explained in Table 7.2. Fig. 7.1 also contains a short description of the function of each unit using the respective energy conversions and technical specifications as to limits of capacity. The flow chart also shows that part of the process and space heat is returned from production to the power station. The amount thus returned is al-

Table 7.2. Problem variables

r_{01}	[t]	surface water	zv_{41}	[t]	10 bar steam for reduction in V4
r_{02}	[t]	mains water	x_{41}	[t]	10 bar process heat from T1
ze_{11}	[t]	treated, desalinated water	x_{42}	[t]	10 bar process heat from T2
ze_{12}	[t]	condensate	x_{43}	[t]	10 bar process heat gained from reduction
zv_{11}	[t]	water for steam generation	x_{51}	[t]	1.5 bar heat from T1
zv_{12}	[t]	water for V1	x_{52}	[t]	1.5 bar heat from T2
zv_{13}	[t]	water for V2	x_{53}	[t]	1.5 bar heat from pressure reduction
zv_{14}	[t]	water for V3	x_{61}	[t]	electricity from T1
zv_{15}	[t]	water for V4	x_{62}	[t]	electricity from T2
ze_{21}	[t]	64 bar steam from B1	x_{63}	[t]	electricity supplied from outside
zv_{21}	[t]	64 bar steam for reduction in V1			
zv_{22}	[t]	64 bar steam for reduction in V2			
zv_{23}	[t]	64 bar steam for operating T1			
ze_{31}	[t]	32 bar steam from B2			
ze_{32}	[t]	32 bar steam reduced in V1			
zv_{31}	[t]	32 bar steam for reduction in V3			
zv_{32}	[t]	32 bar steam for operating T2			
ze_{41}	[t]	10 bar steam from B3			
ze_{42}	[t]	10 bar steam reduced in V2			
ze_{43}	[t]	10 bar steam reduced in V3			

Table 7.3. Production and exploitation coefficients for the power station units

Water treatment unit		
1.035	[t/t]	Ratio of surface/mains water input to desalinated water output
Boilers B1, B2, and B3		
1.0	[t/t]	Ratio of desalinated water input to steam output
Throttle valve V1		
0.9131	[t/t]	Ratio of 64 bar steam input to 32 bar steam output
0.0869	[t/t]	Ratio of added water input to 32 bar steam output
Throttle valve V2		
0.83659	[t/t]	Ratio of 64 bar steam input to 10 bar steam output
0.16341	[t/t]	Ratio of added water input to 10 bar steam output
Throttle valve V3		
0.92963	[t/t]	Ratio of 32 bar steam input to 10 bar steam output
0.07037	[t/t]	Ratio of added water input to 10 bar steam output
Throttle valve V4		
0.94113	[t/t]	Ratio of 10 bar steam input to 1.5 bar steam output
0.05887	[t/t]	Ratio of added water input to 1.5 bar steam output
Turbine T1		
0.3	[t/t]	Ratio of 10 bar steam output to 64 bar steam input
0.7	[t/t]	Ratio of 1.5 bar steam output to 64 bar steam input
133	[kWh/t]	Ratio of electricity output to 64 bar steam input
Turbine T2		
0.3	[t/t]	Ratio of 10 bar steam output to 32 bar steam input
0.7	[t/t]	Ratio of 1.5 bar steam output to 32 bar steam input
65	[kWh/t]	Ratio of electricity output to 32 bar steam input

ways dependent upon the actual use of the heat; for the purpose of this case, however, the amount is given as an average value. The amount of heat which is not recycled as condensate is lost in the production process.

The efficiencies of the different energy resources are dependent on how they are used. The corresponding production and exploitation coefficients of the various units can be seen in Table 7.3. Not all process inputs are expressly mentioned. Insofar as the various resources can be assigned directly to the units, e.g. fuel oil for use in the boilers, or the necessary man-hours, they are immediately converted and registered as fixed or variable output operating costs.

Pollution emission standards prevent the combining of all technical process variations in the power station here under scrutiny. Although the throttle valves and the turbines are not subject to any restrictions other than those of capacity, there are special regulations which have to be observed when using the boilers. In particular, the use of boilers B2 and B3 is ruled out while boiler B1 is in operation. Boiler B2 and B3 can, however, be operated at the same time.

III. Costs

Under the prevailing circumstances it is important that the power station provides the end products in the required amounts, and at a minimum cost.

Table 7.4. Variable costs and overheads

Variable costs

0.19 DM/t	Surface water
2.77 DM/t	Mains water
0.80 DM/t	Recycled condensate
0.096 DM	kWh price for outside electricity in the early shift
0.093 DM	kWh price for outside electricity in the late shift
0.085 DM	kWh price for outside electricity in the night shift
277.08 DM/kW × year	Contract price for standby demand of outside electricity[a]
3.42 DM/t	Water desalination
48 DM/t	Steam generation with boiler B1
47 DM/t	Steam generation with boiler B2
38 DM/t	Steam generation with boiler B3

Overheads[b]

90,890 DM/year	Water treatment unit
699,000 DM/year	Steam generation[c] when B1 is used
649,000 DM/year	Steam generation[c] when B2 is used
649,000 DM/year	Steam generation[c] when B2 plus B3 is used
433,000 DM/year	Steam generation[c] when B3 is used
290,440 DM/year	Turbine T1
213,600 DM/year	Turbine T2

[a] The contract price falls due for the average of the maximum peak requirements in two different months, with a minimum demand of 4,200 kW. This supply basis is symbolized as $\hat{x}_6$.

[b] Staff costs are contained in the overheads.

[c] Cost differences between steam generation variants are exclusively defined by different staff costs.

Costs incurred include both fixed and variable components. Variable costs are essentially the costs of the water used, the electricity supplied from outside, and the process costs of the water treatment and the steam and electricity generation. The overheads are mainly the result of the depreciation and maintenance of the units. Added to these are staff costs closely connected to the operation of certain units.

In Table 7.4 the results of the survey of costs are summarized. The effective tariffs, where they existed, were determined as resource costs. The costs caused by purifying the surface water and the condensing costs of the recycled condensate were determined approximately. The other variable costs contain almost exclusively the tariffs of the residual energy resources used in the process, with regard to the generation of steam this means above all the price of the fuel oil. The staff costs which were incurred were included in the overheads. However, because the need for staff varies according to how the boilers are used, it was necessary to form a set of overheads for each alternative of steam production. The remaining overheads, i.e. the linear depreciation of equipment, are constant, so that the differences shown in Table 7.4 relating to the overheads are caused solely by the variations in the staff required. The alternative of not using a single boiler, and thus lowering overheads still further, was not listed, because even if electricity was only supplied from outside the demand for heat would have to be satisfied by the self generation of steam.

With regard to the policy on an exact hourly basis a suitable distribution of the yearly overheads is necessary. In view of the above explanation a linear distribution on to the working-hours in the power station would appear to be logical. For the study period a total of 6,500 working-hours, distributed equally over the three shifts and all twelve months, was determined.

Total costs are the sum of all the cost components. They have to be minimized with due regard to the given peak hourly demand. It is practical to do this by means of the optimization of the utilized technology. The hourly values for the different periods then give the annual costs of the power station.

IV. The Optimization Model

The problem of finding the best policy for the power station so as to supply all needs can now be formulated as an optimization model. By using the symbols shown in Table 7.2 the following program for an optimal precise hourly assessment can be formulated:

minimize

$$C = 0.19\, r_{01} + 2.77\, r_{02} + 3.42\, ze_{11} + 0.80\, ze_{12} + 48\, ze_{21} + 47\, ze_{31} + 38\, ze_{41} + q^{A}_{63} x_{63}$$

$$\left\{ \begin{array}{l} + 1/6500 \cdot [90{,}890 + 699{,}000\, u_1 + 649{,}000\, u_2 \\ + 433{,}000\, \tilde{u}_3 + 290{,}440 + 213{,}600 + 227.08\, \hat{x}_6] \end{array} \right\}$$

subject to

(1) $r_{01}+r_{02}=1.035\,ze_{11}$,

(2) $zv_{11}=ze_{21}+ze_{31}+ze_{41}$,

(3) a) $zv_{12}=0.08690\,ze_{32}$, b) $zv_{21}=0.91310\,ze_{32}$,

(4) a) $zv_{13}=0.16341\,ze_{42}$, b) $zv_{22}=0.83659\,ze_{42}$,

(5) a) $zv_{14}=0.07037\,ze_{43}$, b) $zv_{31}=0.92963\,ze_{43}$,

(6) a) $zv_{15}=0.05887\,x_{53}$, b) $zv_{41}=0.94113\,x_{53}$,

(7) a) $x_{41}=0.3\,zv_{23}$, b) $x_{51}=0.7\,zv_{23}$, c) $x_{61}=133\,zv_{23}$,

(8) a) $x_{42}=0.3\,zv_{32}$, b) $x_{52}=0.7\,zv_{32}$, c) $x_{62}=65\,zv_{32}$,

(9) $ze_{11}+ze_{12}=zv_{11}+zv_{12}+zv_{13}+zv_{14}+zv_{15}$,

(10) $ze_{21}=zv_{21}+zv_{22}+zv_{23}$,

(11) $ze_{31}+ze_{32}=zv_{31}+zv_{32}$,

(12) $ze_{41}+ze_{42}+ze_{43}=zv_{41}+x_{43}$,

(13) $ze_{11}\leq 30$,

(14) $ze_{12}\leq 0.25\bar{x}_4+0.89\bar{x}_5$,

(15) $ze_{21}\leq 45\,u_1$,

(16) $ze_{31}\leq 28\,u_2$,

(17) $ze_{32}\leq 25$,

(18) $ze_{41}\leq 14\,u_3$,

(19) $ze_{42}\leq 45$,

(20) $ze_{43}\leq 25$,

(21) $x_{53}\leq 18$,

(22) $zv_{23}\leq 45$,

(23) $zv_{32}\leq 23$,

(24) $x_{41}+x_{42}+x_{43}=\bar{x}_4$,

(25) $x_{51}+x_{52}+x_{53}=\bar{x}_5$,

(26) $x_{61}+x_{62}+x_{63}=\bar{x}_6$,

(27) $u_1+u_2+\tilde{u}_3\leq 1$,

(28) $\tilde{u}_3\leq u_3\leq \tilde{u}_3+u_2$,

(29) $u_1,\ u_2,\ u_3,\ \tilde{u}_3\in\{0,\ 1\}$,

(30) Non-negativity of all other variables .

Table 7.5. Process variants in steam generation

Feasible process variants	u_1	u_2	u_3	$\tilde{u}_3$	A priori restrictions on boilers
(1)	1	0	0	0	B1 only
(2)	0	1	1	0	B2 and B3
(3)	0	1	0	0	B2 only
(4)	0	0	1	1	B3 only

To begin with, all types of cost which have any contact with the valuation of the energy processing are taken into account in the objective function. A linear distribution took place of the annual costs on to the 6,500 working hours. In each of the assessment periods under review the current kilowatt price for electricity supplied from outside is used for q_{63}^{A}.

In order to avoid burdening the model calculation with integer variables, the feasible process variants for the generation of steam were subjected to a separate examination from the beginning. Table 7.5 gives a review of the corresponding differentiation. In addition, it will be seen that the generally permitted variant (3) is dominated by (2) because of the lower production capacity at constant capacity costs. This means that although it can be economic to operate only boiler B2 on a periodic basis, a corresponding a priori restriction does not make sense. For this reason, only the variants (1), (2), and (4) were tested for each survey period.

The contract demand base $\hat{x}_6$ for the delivery price of that part of the electricity coming from outside was assumed as a non-decisive factor in the comparison of processes on a periodic basis as $\hat{x}_6$ has to be calculated on a maximum over several periods, and the peak purchasing values are only available after completion of all the periodic calculations. This does not of course apply to the final assessment of the whole year's policy. Because of the cost-efficiency of maximum purchasing of electricity, retroactive decisions as to the order in which the variants were rated cannot be ruled out and have to be taken into account.

If the model is used as described there are two immediate consequences. Firstly, the approach is reduced to linear programming submodels without integer variables, the parameters of which vary in the manner discussed, so that several calculations have to be carried out and the results compared. Secondly, that part of costs shown in curved brackets takes on the character of fixed overheads and does not enter the optimization, although they are included in the model runs for completeness.

The constraints (1) – (8) describe the energy conversion processes in the individual units, that is, they display their linear limitational production functions. All these equations remain constant for all calculation variants. The same applies to the balance equations (9) – (12), which connect the generated and consumed quantities of two consecutive production levels with one another, i.e. there occur no energy losses within the power station. In addition, the conditions (13) – (23) and (24) – (26) introduce the capacities of the units and the periodic energy demand respectively into the formulation of the problem. Capacity restrictions for the boilers have to be adjusted for each process variant as it is tested (see Table 7.5). Conditions (27) – (29) have already been resolved, thus leaving the non-negativity conditions (30) as the sole technical model restrictions.

The derived linear programming problems show, apart from the non-negativity conditions, only 29 continuous variables and 34 restrictions. Problems of this magnitude can be solved quickly using standard software and a personal computer.

V. Some Results

The essential results of the hourly optimization can be seen in Table 7.6. The first column contains, along with the value of the objective function, the variables for electricity, 10 bar process heat, and 1.5 bar heat differentiated as to their origins in the generation process. In the following columns the periodic calculations are compiled within the periods divided as to the different alternatives of steam generation. It can be seen that the technologically feasible variant (4) does not result in a solution for certain demand values, as the capacity of the boiler B3 is not sufficient in these cases to satisfy the given demand for heat.

As far as the periodic differentiation of the analysis is concerned, it is worth noting that in 6 out of 9 survey sections the most satisfactory generation of energy in terms of cost was achieved mainly by using boiler B1 and turbine T1. A rigid joint production using T1 leads automatically, however, to the situation that the total demand for space heating cannot be satisfied using T1, as the ratio of the energy exploitation does not correspond to the ratio of the demand quantities. In the present case the demand for 10 bar steam cannot be fully satisfied, as surplus production from T1 would lead to the inadmissible pass-out of 1.5 bar steam. Because of this, the demand for 10 bar heat is partially met (in period VII more than 50%) through the throttling of the original 64 bar steam by feeding in desalinated water. This causes, however, a simultaneous reduction in the amount of steam available for the generation of electricity. If the environmental regulations were relaxed so as to make a surplus production of 1.5 bar steam possible, or to enable steam to be eliminated or condensated at reasonable costs, T1 could then definitely be employed economically to full capacity. Under the prevailing circumstances the turbine's maximum capacity of 5,985 kWh can only be used to 44% (periods VIII and IX). The remainder of the required electricity has to be dispatched from outside.

An exact hourly observation leads to the recommendation that, for reasons of economy, the smallest boiler B3 should be utilized in the three periods IV – VI, as these are the summer months July and August in which the demand for space and process heat is comparatively low. In period IV boiler B2 is to be used as a back up for the steam generation, while B1 does not come into service at all. Because B3 can only generate 10 bar steam this policy means that at the same time almost all the necessary electricity has to be supplied from outside. The power station's activities are reduced to the minimum of self-supply of heating and process heat.

Table 7.6. Results on the basis of an hourly calculation

	Period I (17, 24, 8600)*			Period II (14, 21, 8600)			Period III (14, 20, 6500)			Period IV (10, 6, 7700)			Period V (8, 5, 7700)		
	(1)	(2)	(4)**	(1)	(2)	(4)	(1)	(2)	(4)	(1)	(2)	(4)	(1)	(2)	(4)
C	[2867]	3005		[2621]	2720		[2370]	2441		1863	[1857]		1722	1707	[1674]
x_{41}	10.29	–		9	–		8.57	–		2.57	–		2.41	–	–
x_{42}	–	6.9		–	6.18		–	5.89		–	0.52		–	–	–
x_{43}	6.71	10.1		5	7.82		5.43	8.11		7.43	9.48		5.86	8	8
x_{51}	24	–	no solution	21	–	no solution	20	–	no solution	6	–	no solution	5	–	–
x_{52}	–	16.1		–	14.43		–	13.74		–	1.2		–	–	–
x_{53}	–	7.9		–	6.57		–	6.26		–	4.8		–	5	5
x_{61}	4560	–		3990	–		3800	–		1140	–		950	–	–
x_{62}	–	1495		–	1340		–	1276		–	112		–	–	–
x_{63}	4040	7105		4610	7260		2700	5224		6560	7588		6750	7700	7700

	Period VI (7, 5, 6900)			Period VII (13, 15, 8300)			Period VIII (10, 14, 8300)			Period IX (10, 14, 7200)		
	(1)	(2)	(4)	(1)	(2)	(4)	(1)	(2)	(4)	(1)	(2)	(4)
C	1557	1537	[1504]	[2358]	2421		[2178]	2220		[2039]	2065	
x_{41}	2.14	–	–	6.43	–		6	–		6	–	
x_{42}	–	–	–	–	4.1		–	2.87		–	2.87	
x_{43}	4.86	7	7	6.57	8.9		4	7.13		4	7.13	
x_{51}	5	–	–	15	–	no solution	14	–	no solution	14	–	no solution
x_{52}	–	–	–	–	9.58		–	6.70		–	6.7	
x_{53}	–	5	5	–	5.42		–	7.30		–	7.3	
x_{61}	950	–	–	2850	–		2660	–		2660	–	
x_{62}	–	–	–	–	889		–	622		–	622	
x_{63}	5950	6900	6900	5450	7411		5640	7678		4540	6578	

* Abbreviation: supply vectors ($\bar{x}_4$, $\bar{x}_5$, $\bar{x}_6$)

** Boiler utilization variants from Table 7.5

To ascertain the costs the contract price of outside electricity was based on a constant amount of 6,750 kW. The minimum hourly total costs for each period are bracketed.

The analysis for the whole year demonstrates, however, through the concrete requirement figures, that the simple addition of the optimum results in each period does not lead to an overall optimal policy. Merely the fact that for July and August a policy existed of obtaining the total amount of electricity from outside would have a major influence on the contract price for the whole year. This spillovers must be subsequently measured and assessed. Compared with the solution derived above a permanent utilization of boiler B1 even in the summer months is successful in reducing the costs of outside electricity, but at the expense of an increase in running costs in the periods IV – VI. This offset in costs is effective inasmuch as the advantages of a lower electricity contract throughout the whole year outweigh the increase in the power station's operating costs in the summer months. The savings amount to about 225,000 DM; in view of the annual costs of 15 million DM this means a saving of 1.5%.

A thorough-going recommendation of a strategy for the utilization of the power station would have consisted ultimately, given a priori knowledge of the demand figures used in the calculations, in electricity generation with turbine T1 being pushed permanently as far as possible and using outside supplies only in cases of technologically caused shortages. This is justified by the fact that, on the one hand, self generation is fundamentally advantageous as compared with outside supply because of the existing oil and electricity prices and the more favourable technological alternatives. On the other hand, in a comparison of these technological possibilities in a narrower sense, 64 bar steam provides the best efficiencies as regards the generation of electricity and heat, an advantage which justifies the insignificant increases in oil and personnel cost as compared with the other alternatives. The amount of water necessary remains irrelevant as regards the utilization of boilers, because the degree of effectiveness is constant for all process variants – one ton of water provides one ton of steam. Instead of desalinated surface or mains water the cheaper condensate, recycled from the manufacturing process, is to be preferred for the generation of steam. Only the residual requirements need then be treated.

VI. Subjects of Further Research

The aforesaid practical study was occasioned by the general, nonspecific uncertainty in the firm as regards the economy of its own power station, which was planned and constructed under completely different environmental conditions. Changes in the prices of raw materials, especially fuel oil, and pollution emission standards were at that time either not foreseeable or played a minor part. The economic analysis, therefore, had to take on the function of a control test, which was to provide additional information about the flexibility and stability of management policy as pursued up to the present, and to identify any corrections which might prove necessary. A further aim of the study was to recognize, as exactly as possible, the structure of the multi-stage production problem for processing various energy products, and to use this to propose instruments for a systematic method of assessment and planning, which would generally be of assistance in decision support.

In order to find a practical solution precisely for this second object it is necessary to extend the problem's horizon, starting directly with the chosen formulation of the problem at hand. There are first of all fluctuations in demand for the individual types of energy and developments in the resource prices, unforeseeable when the yearly plans were being made, which necessitate a sensitivity analysis of the proposed solution. For example, the break-even point for electricity supplied from outside as compared with self generation using turbine T1 is roughly 7.22 Pfg/kWh. In comparison, the overall delivery price of electricity at present totals 14.04 Pfg/kWh.

Fluctuations in demand within the plant can be allowed for with greater accuracy by coupling the decisions concerning the energy generation more closely to the production planning in the firm as a whole. WOLF (1987) provides information about this possibility.

As far as long-term strategy for the utilization of the power station is concerned there are several investment possibilities which present themselves after taking stock of the whole situation, and which can certainly be of service as regards the optimal policy. Firstly, it is worth checking the capacities of future replacement units. In view of the fact, for example, that the average load of the boiler B1 is only about 60% and that its maximum load (in period I) does not reach 90%, and that boilers B2 and B3, and turbine T2, are mainly only used during breakdowns of B1 and T1, the question can be posed as to whether breakdown probabilities and environmental uncertainties justify organizational slack on such a large scale. Secondly, an increase in investment can also be considered as a means towards long-term optimal policy. For example, policies based on the profitable use of electricity will require the installation of extra condensers to enable the surplus generation of steam in the first place. If, beyond this, it is intended to make the power station economically less dependent, for example by organizing it as a profit or investment center, thus automatically increasing the number of market transactions, then the considerations regarding investment which were mentioned at the outset once again come to the forefront.

References

Albach, H.: Produktionsplanung auf der Grundlage technischer Verbrauchsfunktionen, in: Brandt, L. (ed.): Arbeitsgemeinschaft für Forschung des Landes Nordrhein-Westfalen, No. 105, Köln-Opladen 1962, pp. 45–98.

Constantopoulos, P., Larson, R.C., and Schweppe, F.C.: Decision Models for Electric Load Management by Consumers Facing a Variable Price of Electricity, in: Lev, B. (ed.): Energy Models and Studies, Amsterdam-New York 1983, pp. 273–291.

Fatti, L.P.: Optimal Smoothing of Demand for Industrial Gas, in: The Journal of the Operational Research Society, 1983, pp. 583–590.

Porter, R.W., and Mastanaiah, K.: Thermal-Economic Analysis of Heat-Matched Industrial Cogeneration Systems, in: Energy, 1982, pp. 171–187.

Skea, J.F.: Switching from Oil to Coal Firing for Steam Raising, in: Energy Policy, 1981, pp. 205–215.

Stocks, K.J., Maher, K.J., Le, D., and Bannister, Ch.: An LP Model for Assessing Cogeneration Strategies, in: Omega, 1985, pp. 541–554.

Wolf, J.: Industriebetriebliche Energienutzung und Produktionsplanung, Heidelberg-Wien 1987.

Chapter 8

Decision Support for Integrated Production Planning A Case Study from Chemical Industry

MANFRED GRAUER

I. Introduction

Economic as well as technological considerations have dominated the evolution of industrial structures. These factors have been treated extensively but mostly separately by economists and engineers in the past. However, another major factor which has begun to have a decisive influence on the performance of industrial processes is technological risk as well as public and environmental health considerations, in particular those related to toxic and hazardous substances used in industrial production processes.

The issues of controlling process risk, waste streams, and potential environmental consequences of accidental or routine release of hazardous substances are rapidly gaining in importance as a non-negligible part in production planning. To characterize the order of magnitude of those waste streams from a technological and economic point of view it seems worth to mention that they are estimated in WORLD RESOURCES (1987) for the year 1986 by 330 million metric tons. These developments and the availability of advanced information processing tools have forced to take into account all these criteria (e.g. technological, economical, ecological) and to integrate the stages of the engineering design of production processes, with its control and the corresponding production planning problems in normal and abnormal situations.

The paper tries to approach some of the problems under the above mentioned multiple criteria and to indicate the role of decision support for computer-aided integrated production planning. This will be done analysing a production process from chemical industry under normal and exceptional circumstances (emergency situations).

II. The Problem

The experiences in developing and using optimization-based decision support (see e.g. GRAUER 1983) were the basis to concentrate on integrated production planning under the prospect of hazardous byproducts. On the other hand an ongoing project at the International Institute for Applied Systems Analysis (IIASA) in Laxenburg/Austria was the initial point to choose the chemical process of direct chlorination of phenol as a case (see FEDRA 1985 and GRAUER and FEDRA 1987). This process raised attention in connection with the Seveso-

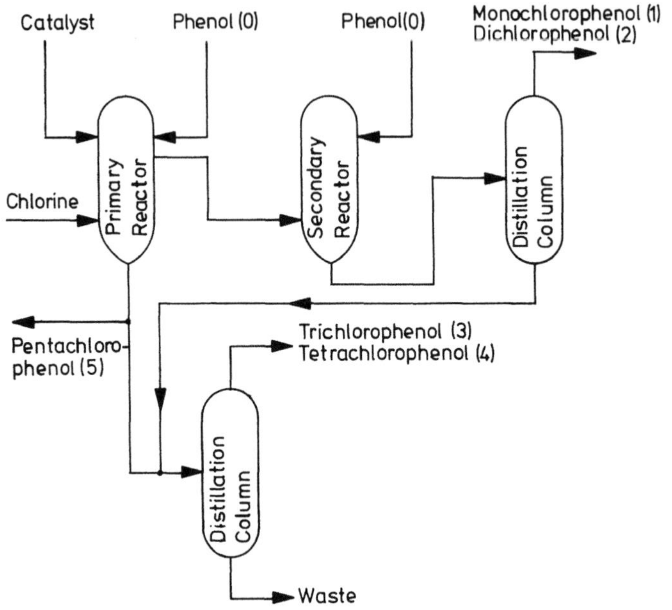

Fig. 8.1. Simplified technological schema of the chemical plant (direct chlorination of phenol)

accident in Italy in 1976. A simplified technological schema of that process is given in Fig. 8.1.

The extremly toxic 2,3,7,8-tetrachlorodibenzo-p-dioxin and 2,3,4,7-pentachlorodibenzofuran are byproducts of that process. The process representation as well as the process risk description is based on GOLDFARB et al. (1981) and GRAUER et al. (1986). Thereby it is assumed that the chlorination of phenol (0) proceeds stepwise so that five chlorophenols of commercial value are obtained: monochlorophenol (1), dichlorophenol (2), trichlorophenol (3), tetrachlorophenol (4), and pentachlorophenol (5). These chlorophenols are used as herbicides, bactericides, fungicides, and chemical intermediates.

The subprocesses and substances involved in that manufacturing are summarized below in some more detail:

phenol (0) + chlorine → monochlorophenol (1) + HCl ,

monochlorophenol(1) + chlorine → 2,4-dichlorophenol (2) + HCl ,

2,4-dichlorophenol (2) + chlorine + ($AlCl_3$) → 2,4,6-trichlorophenol (3) + HCl ,

2,4,6-trichlorophenol (3) + chlorine + ($AlCl_3$) → 2,3,4,6-tetrachlorophenol (4) + HCl ,

2,3,4,6-tetrachlorophenol (4) + chlorine + ($AlCl_3$) → pentachlorophenol (5) + HCl ,

pentachlorophenol (5) + chlorine + ($AlCl_3$) → dioxin (6) .

The numbers in parenthesis are used as abbreviations of the substances for the future use in the paper, and $AlCl_3$ indicates a catalyst (anhydrous aluminium chloride) which is added to the reactor when the chlorination has proceeded to the dichlorophenol stage (see Fig. 8.1).

The reaction starts at a temperature of about 70 °C and is increased up to 180 °C to maintain a differential temperature of 10 °C over the product melting point. The process is running batchwise with a total period of chlorination of generally 8 to 10 hours. As shown in Fig. 8.1, phenol is added to the primary reactor and to the secondary reactor. The reactors are heated and the temperature profil controls the production process. The process starts with the introduction of chlorine into the first reactor. The offgases from the first reactor are passed into the secondary one. Separation units, mainly consisting of distillation columns, allow the extraction of the desired products and the hazardous by-product, here for simplicity assumed to be only dioxin.

The problem of control and management of this chemical plant consists of meeting the demand of the five products (pentachlorophenol is the main product, but mono-, di-, tri- and tetrachlorophenol are also products), and/or maximizing the target products in a time-minimal production period. Here the interdependencies between the different outputs of the process have to be taken into account. That means for instance that the amount of pentachlorophenol is a result of all steps of the reaction and any intermediate product would reduce its throughput. This requests to understand the production profil as a compromise of all target products which meets best the current demand profil.

Moreover to these production criteria the amount of toxic substances (e.g. dioxin) has to be minimized or at least to keep below certain upper bounds. This leads in the normal production mode as well as in non-normal cases (e.g. runaway reaction, operator error, equipment failure, fire, explosion etc.) to the necessity to deal with a dynamic multiple-criteria decision problem. The use of mathematical models as support in such a decision situation is the subject of the further discussion.

On the other hand these mathematical models have to be used already in the design stage when the decision about the principal process chemistry is made and the engineering design of the layout of the plant is done. Of course the decision variables are then different from those in the case of production planning for a given plant. But as well known, in the design stage about 80% of the economic efficiency is determined and also the possibilities to meet ecological and safety requirements. This calls for an integration of decision support for computer-aided management in all these stages.

III. The Mathematical Model of the Production Process

The composition and physical properties of reacting materials influence stability, efficiency, and quality when manufacturing chemicals. In large-scale operations as indicated above the main reaction is often influenced by other reactions and/or perturbed by less understood secondary reactions. An analytical model (a deep model in the terminology of knowledge processing) can only be an approximation of the production process under analysis and has to be supported by other information and knowledge (shallow model).

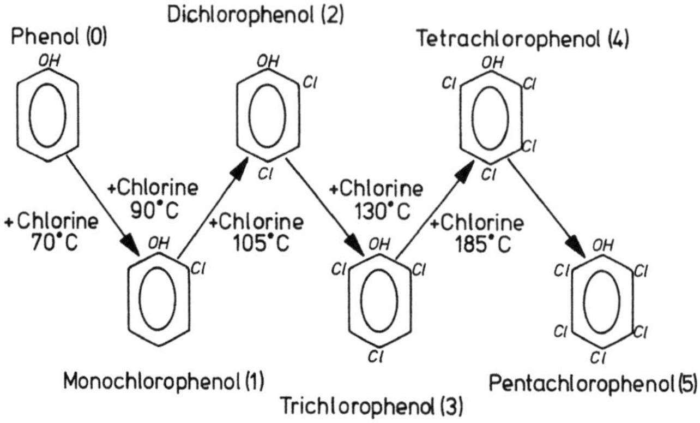

Fig. 8.2. Direct chlorination of phenol (simplified process chemistry)

For the chlorination process the theoretically and experimentally verified conceptual model (deep) is that of a consecutive reaction from phenol (0) over pentachlorophenol (5) to dioxin (6) of the following type:

$$(0) \rightarrow (1) \rightarrow (2) \rightarrow (3) \rightarrow (4) \rightarrow (5) \rightarrow (6) \quad .$$

This concept implies ideal mixture conditions in the reactor. The chemistry underlying this concept is presented in Fig. 8.2 in a simplified manner.

The mathematical formulation of this concept is a system of ten differential equations (DE) consisting of (for details see GRAUER et al. 1986):

7 DE for the mass-balances of all substances 0 through 6,
1 DE for the mass-balance of chlorine in the liquid phase,
1 DE for the mass-balance of chlorine-consumption and
1 DE for the energy-balance in the reactor.

In addition to the system of differential equations a set of highly nonlinear algebraic equations represents the interdependencies of densities, reaction coefficients, heat transfer coefficients, etc. from time, temperature, pressure, and material composition, e.g. concentration.

Systems of differential equations of the above described type are usually "stiff", e.g. their solutions contain parts with rapidly decaying components. They are important in numerical analysis because they frequently arise in practical problems (see SMITH 1987) and because they are difficult to solve by traditional numerical methods. For the above system the method of implicit backward differentiation formulas is used (see e.g. SMITH 1987).

The mathematical description of the process is presented here in such detail to give an understanding of the constraints for the finally to solve control/optimization task as part of a decision problem.

IV. Formulation of the Production Control Problem

The mathematical model to describe the dynamic behavior of such continuous or semicontinuous (batch-wise) production processes as given in the previous Section III is usually a system of differential equations as follows:

(1) $$\frac{dx(t)}{dt} = F[x(t), u(t), t] \ , \qquad x(t) \in \mathbb{R}^m \ , \qquad u(t) \in \mathbb{R}^n$$

with the given initial state of the production system at the time $t = t_0$:

(2) $$x(t_0) = x_0 \ , \qquad t \in [t_0, t_f] \ ,$$

where x is the state and u the control variable and $[t_0, t_f]$ indicate the planning period.

Usually the state- and the control-variables are subject to further constraints:

(3) $$g[x(t), u(t), t] \leq 0 \ ,$$

and

(4) $$h[x(t), u(t), t] = 0 \ .$$

It is assumed that the performance measure has the form of the following scalar-valued functional:

(5) $$\min \left\{ \int_{t_0}^{t_f} f[x(t), u(t), t] \, dt + \Phi[x(t_f)] \right\} \ .$$

The functional (5) represents the time-dependence of the criteria for the decision processes like economic ones or safety considerations. Under normal production conditions this can be an overall cost-function. For decision problems connected with abnormal situations the integral can represent the amount of toxic substances for which the danger exists to be accidentically released and which has to be minimized.

The most common approach to solve this class of problems is to convert such an optimal control problem (1) – (5) into a problem which can be solved by nonlinear programming methods. This can be done by approximating the continuous time-function $u(t)$, i.e. the control-trajectory (or policy) by a discrete time-function. The new decision trajectory or decision policy is derived by choosing an appropriate structure with finitely many free parameters and to consider the state variables $x(t)$ as functions of the control variables through the differential equations (1). Hereby the problem is solved in the reduced space of controls, and the state variables are obtained by the solution of an initial value problem.

Following this approach the interval $[t_0, t_f]$ is partitioned into a (not necessarily equidistant) time-grid:

(6) $$t_0 = t_1 < t_2 < \ldots < t_j < \ldots < t_k = t_f \ ,$$

and on this grid the finite set of control variables is defined:

(7) $$u(t) = [u_1(t_1), \ \ldots, u_n(t_1), \ u_1(t_2), \ \ldots, \ u_n(t_{k-1})] \ .$$

The infinite control or decision problem (1) – (5) can then be restated in the following way:

(8) $$\frac{dx(t)}{dt} = F[x(t), u_i(t_j), t] \ , \quad i = 1, \ldots, n \ , \quad j = 1, \ldots, k-1 \ ,$$

with

(9) $$x(t_0) = x_0 \ , \quad t \in [t_0, t_f] \ ,$$

(10) $$g[x(t), u_i(t_j), t] \leq 0 \ , \quad i = 1, \ldots, n \ , \quad j = 1, \ldots, k-1 \ ,$$

(11) $$h[x(t), u_i(t_j), t] = 0 \ , \quad i = 1, \ldots, n \ , \quad j = 1, \ldots, k-1 \ ,$$

and

(12) $$\min \left\{ \int_{t_0}^{t_f} f[x(t), u(t), t] \, dt + \Phi[x(t_f)] \right\} .$$

The nonlinear single-criterion programming problem then is to find the piecewise-constant control strategy (7) which minimizes (12) under the constraints (8) – (11). This is a nonlinear programming problem with an iterative procedure in the constraints and in the objective function. To solve this problem in a reliable manner and sufficiently fast only the direct search (derivative free) Polytop-method was able. This algorithm is from a library of standard programming methods (GRAUER 1987).

Next a brief description of the approach for the multiple-criteria production planning problem actually to be solved is given.

V. The Multiple-Criteria Approach

Production planning problems as stated earlier are multiple-criteria ones. An approach is needed which transforms the originally multiple-criteria problem into a scalar-valued one to analyse it. Such a transformation has to have two main features. First it has to represent in some way the preference structure of the user to allow to support the decision making process and secondly it must be numerically efficient.

For this transformation the idea of satisficing behavior and its mathematical formalization for decision support is used as developed by WIERZBICKI (1982). To give a brief understanding of this approach for the purposes of the current analysis it is assumed that the decision space $E_u = \mathbb{R}^n$ and the set of admissible decisions $U_0 \subset \mathbb{R}^n$ are compact. Let there be p criteria or objectives of interest to the decision maker, $E_q = \mathbb{R}^p$, and let the outcome mapping $f\colon U_0 \to \mathbb{R}^p$ be continuous, hence the set of attainable outcomes $Q_0 = f(U_0)$ be also compact. Under the assumption that the decision maker wants to maximize all outcomes, the partial ordering of the outcome space is implied by the positive cone $D = \mathbb{R}^p$. In WIERZBICKI (1982) is stated if the cone D is closed and the set Q_0 is compact, then there exist D-efficient or D-optimal elements of Q_0. These are such elements $\hat{q} \in Q_0$ that

$$Q_0 \cap (\hat{q} + \tilde{D}) = \phi \ ,$$

where $\tilde{D}$ is the cone D except its origin and $\hat{q}+\tilde{D}$ denotes the set $\tilde{D}$ shifted by $\hat{q}$. The corresponding decisions $\hat{u} \in U_0$ such that $\hat{q}=f(\hat{u})$ are D-efficient as well. The set of all D-efficient outcomes is then:

$$\hat{Q}_0 = \{\hat{q} \in Q_0 \colon Q_0 \cap (\hat{q}+\tilde{D}) = \emptyset\} \ .$$

For applications in decision support systems (see e.g. GRAUER 1983) a substitute scalarizing function is introduced. The so-called achievement function is a transformation from the p-dimensional objective space into the one-dimensional space: $s\colon \mathbb{R}^p \to \mathbb{R}^1$. This function is guided by reference levels or aspirations of the decision maker about target values or trajectories for the criteria $(\bar{q}_i)$ and its maximization leads to D-efficient decisions. In the multiple-criteria dynamic case as discussed here this function can be written as (see GRAUER 1983):

$$(13) \quad s(q,\bar{q}) = \min_{t_1 \le t \le t_k} \ \min_{1 \le i \le p} \mu_i[q_i(t)-\bar{q}_i(t)] + \ldots + \delta \sum_{t=t_0}^{t_f} \sum_{i=1}^{p} \mu_i[q_i(t)-\bar{q}_i(t)] \ ,$$

where $\delta > 1, \mu_i \ge 0$ and $\sum_{i=1}^{p} \mu_i = 1$.

The solution of the multiple-criteria problem consists now in finding those decisions (7) which maximize (13) for given reference levels $\bar{q}_i$ under the constraints (8) – (11). This is in principal the same numerical problem as stated in the previous section, i.e. a nonlinear programming problem with an iterative procedure (the solution of the system of differential equations) in the constraints and in the objective function. For the numerical solution the same statements as in Section IV are valid.

VI. A Concept for the Use of the Model in Decision Support

The current prototype module of the mathematical model could support a single user in analysing production planning problems under multiple criteria in an off-line manner. Also in the case of a non-normal situation of the plant the user can run the problem with the objectives either to meet the production goals and/or to satisfy the safety requirements.

A decision support system (DSS) of this off-line type is described in FEDRA (1986) and GRAUER and FEDRA (1987). There the interface to the DSS for this dialog is problem-oriented and the whole system is userdriven. Setting certain target levels for the production, for the production period (t_f), and/or for the minimization of certain hazardous substances, the user will be answered by the system with the corresponding efficient control strategy and the state trajectory of the chemical plant. The Fig. 8.3 shows a presentation of the process dynamics for one specified efficient decision case.

One production planning run for a result as specified in Fig. 8.3 takes on an IMB-PC XT/AT with an APU-8087 about 40 to 50 minutes. This are then about 20 to 30 function evaluations for the underlying nonlinear programming problem. The same problem is solved on a VAX 11/780 in less than 10 minutes and on a SUN 3/260 in less than 30 seconds.

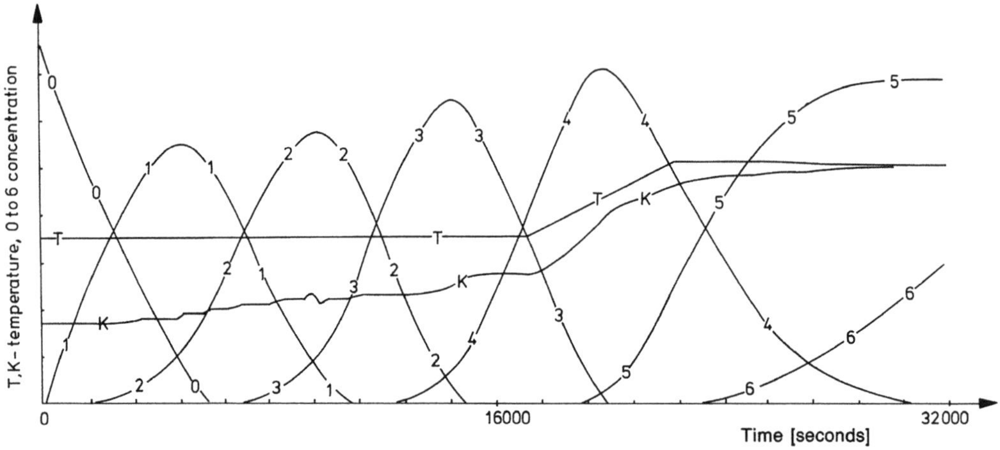

Fig. 8.3. Process dynamics of a chemical plant for the direct chlorination of phenol (T, K-temperature in the reactor and of the coolant in °C and (0), (1), (2), (3), (4), (5), (6)-concentration in kmol/m^3 of phenol, mono-, di-, tri-, tetra-, pentachlorophenol, dioxin) after optimization (efficient decision case)

The software for modelling and optimization is implemented in Fortran-77 and the interface in C. A test to realize both the interface and the numerical part in the language C to improve the portability of the software system is currently underway (GRAUER et al. 1988).

VII. Some Conclusions

The concept of this prototyp module of the mathematical model in that stage is not only usable for the specific process under discussion here, but in a broader range of integrated production planning and the management of emergency

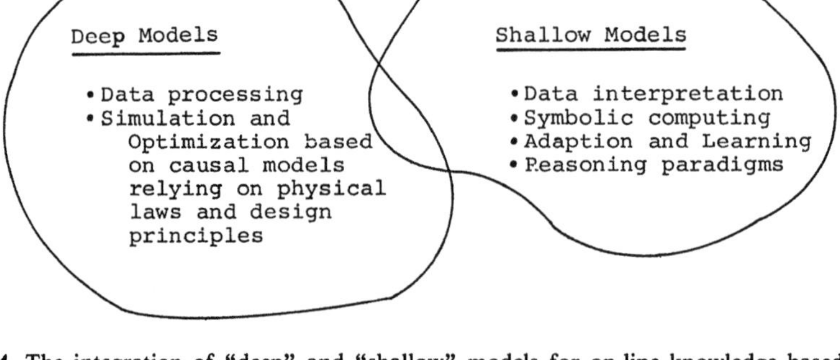

Fig. 8.4. The integration of "deep" and "shallow" models for on-line knowledge based decision support

cases. The use of mathematical models for decision support in emergency situations is already known for high-order corporate crisis management (see e.g. ANDRIOLE 1986) and in industry for the management of nuclear power plants (see e.g. EMBRY and HUMPHREYS 1985). For the application as decision aid in on-line production management the step has to be done from "deep" to "shallow" mathematical models which is indicated in Fig. 8.4. This means the integration of knowledge processing and on-line information from process measurements into DSS. The development of information and communication technology and sensor technology supports this step already and research in this direction is underway (see GALLANTI and GUIDA 1986).

References

Andriole, St.: Microcomputer-based decision support for high-order corporate crisis management, in: Andriole, St. (ed.): Microcomputer decision support systems: design, implementation and evaluation, Amsterdam 1986.

D'Ambrosio, B., Fehling, M.R., Forrest, St., Raulefs, P., and Wilber, B.M.: Real-time process management for materials composition in chemical manufacturing, in: IEEE Expert, Summer 1987, pp. 80–93.

Embry, D., and Humphreys, P.: Support for decision making and problem solving in abnormal conditions in nuclear power plants, in: Methlie, L.B., and Sprague, R.H. (eds.): Knowledge representation for decision support systems, Amsterdam 1985, pp. 109–124.

Fedra, K.: Advanced decision-oriented software for the management of hazardous substances. Part I: structure and design, CP-85-18, International Institute for Applied Systems Analysis, Laxenburg/Austria 1985.

Fedra, K.: Advanced decision-oriented software for the management of hazardous substances. Part II: a demonstration prototyp system, CP-86-10, International Institute for Applied Systems Analysis, Laxenburg/Austria 1986.

Gallanti, M., and Guida, G.: Intelligent decision aids for process environments: an expert system approach, in: Hollnagel, E., et al. (eds.): NATO ASI Series, Vol. F21, Intelligent decision support in process environments, Berlin-Heidelberg-New York 1986.

Goldfarb, A.S., Goldgraben, G.R., Herrick, E.C., Quellette, R.P., and Cheremisinoff, P.N.: Organic Chemicals Manufacturing Hazards, Ann Arbor 1981.

Grauer, M.: A Dynamic Interactive Decision Analysis and Support System (DIDASS) – User's Guide, WP-83-60, International Institute for Applied Systems Analysis, Laxenburg/Austria 1983.

Grauer, M., et al.: Modeling and Optimization of the Process of Chlorination of Phenol, Internal Report of the International Institute for Applied Systems Analysis, Laxenburg/Austria 1986.

Grauer, M.: DSS-toolbox (optimization methods), release 2.d, University of Dortmund, Department of Computer Science 1987.

Grauer, M., and Fedra, K.: Intelligent decision support for technology assessment: the chemical process industry, in: Sawaragi, Y., Inoue, K., and Nakayama, H. (eds.): Toward interactive and intelligent decision support systems, Volume 2, Berlin-Heidelberg-New York 1987, pp. 40–55.

Grauer, M., Albers, St., and Frommberger, M.: Concept of a window-based interface for SIAD (système intégré d'aide a la décision), paper at 27th meeting of the European Working Group on MCDA, Mons 1988.

Smith, J.M.: Mathematical modelling and digital simulation for engineers and scientists, New York 1987.

Wierzbicki, A.: A mathematical basis for satisfying decision making, in: Mathematical Modelling, 1982, pp. 391–405.

World Resources 1987: A report by The International Institute for Environment and Development and The World Ressources Institute, New York 1987.

Chapter 9

Optimal Program Planning in Joint Production

GÜNTER FANDEL

I. Special Features of Program Planning in Joint Production

Program planning in joint production raises a series of additional problems as compared with the standard approach of linear programming which is normally used in multi-stage production (FANDEL 1980, p. 96; KILGER 1973, p. 160). The statically formulated standard approach to the optimal planning of the production program assumes constant rates of profit, fixed production coefficients and defined maximum sales quantities of the various marketable intermediate and final products, and maximizes the sum of the cost covering margins while the production quantities must not exceed the maximum sales quantities. This procedure cannot simply be applied to the situation of production program planning in joint production, thus requiring corresponding modifications.

Apart from usual converging production processes which may occur, joint production is essentially characterized (RIEBEL 1971, p. 733; RIEBEL 1979, col. 1009) by the fact that in addition on the different stages of production there are splitting processes. In the course of these processes, by the processing of certain inputs several by-products will occur simultaneously. For technological reasons, the amounts of these by-products are related to each other. Formally, such joint processes are easily taken into account within the framework of the standard approach by additionally integrating a system of yield constraints which indicate how many units of the various joint products of a splitting process can be obtained from one unit of the input concerned. In this connection final and intermediate products as well as waste products may be regarded as possible joint products. It is relatively simple to extend the standard approach analogously, even if the quantities of the joint products are in no fixed relationships to each other, i.e. there are no rigid proportions, but the latter can vary within certain limits, depending on the splitting process chosen. These cases of loose proportions which, in a way, resemble the problem of choosing the suitable procedure in multi-stage multiple product manufacturing can be coped with by a series of more or less efficient formulation approaches which are available (ALBACH 1962, p. 45; JACOB 1962, p. 205) and can directly be used for the purposes of program planning in joint production processes.

In this connection, however, it is more difficult to deal with the problem of specifying and evaluating the quantities of the final and intermediate products and maintaining the concept of cost covering margin accounting for profit planning. With respect to this problem two different solution approaches are to be found in the literature. One of these approaches is based upon v. STACKELBERG's idea (1932, p. 53) to use the findings from simple production or separate multiple product manufacturing in such a way that those goods which result from the various joint production processes are each combined in product packages. These product packages can then be interpreted as distinguishable individual products. Extending this idea RIEBEL (1970, col. 994, and 1971, p. 741) has developed calculation schemes which can be used to determine cost covering margins of such product packages for the different types of joint production. The dimensions of the methodical and practical difficulties which this procedure might entail, however, are largely dependent upon which relative proportions between the joint products and their progenies have to be considered in the following processing. Thus in multiple joint production processes with combined processing, the problem that product packages must be made within product packages is difficult to be coped with, primarily because of the enormous quantity of data (KRUSCHWITZ 1974, p. 96). Additionally, how the individual maximum sales quantities of the marketable joint products of a product package are to be dealt with in regard to this product bundling and to what extent the determination of the cost covering margin is affected are unresolved questions. On the other hand, the attempt to determine margins even for the individual joint products, would entail the much more serious problem of adequate costs. Since the separability of the cost functions which is required for this determination is contradictory to the joint production, the problem of imputed cost remains unsolvable (KILGER 1973, p. 353).

In order to avoid the above difficulties KILGER (1973, p. 361), BIETHAHN (1974, p. B167), and KRUSCHWITZ (1974, p. 96) made another solution approach which has led to different linear programming approaches for optimal program planning in joint production. These approaches largely follow the process oriented calculation model developed by ALBACH (1962) for problems involving the choice of procedure in multi-stage multiple product manufacturing. Consequently, they start first by listing separately the variable unit costs of the goods produced in the individual processes and further used in joint processes on the one hand, and the selling prices of the final and the marketable intermediate products on the other hand. In a second step, the overall profits are then compared with the total manufacturing cost in the profit function; thus, the problem of separate determination of cost covering margin rates is avoided. In this connection, continuity and balance constraints will guarantee that as much is produced at each manufacturing stage as is required for production at each subsequent stage.

With these modifications the standard approach can largely be adapted to the theoretical and practical requirements of joint production planning. One aspect, however, has so far received little notice in the literature or has been treated in a contradictory way. This concerns the following problem: In non-joint production environments the requirement that the quantities of production must not exceed the maximum sales quantities is a reasonable premise of the standard approach. But to what extent is this requirement also valid for program planning in joint production, should it be restricted or even abolished, and to what extent will the production program and its optimality be affected as a result of such a different treatment? This requirement is certainly reasonable for non-joint production, because production quantities which exceed the maximum sales quantities are no longer remunerative. For further increase in profits the resources claimed by them could be made available to products with positive profit margins the quantities of which have not yet reached the maximum sales quantities. This transfer of resources which depends on the relative profit margins and maximum sales quantities is, however, not possible for the joint products of a splitting process, because the product quantities are automatically obtained in certain ratios. Consequently, the requirement that the quantities of production of the marketable joint products are not to exceed the maximum sales quantities will not automatically comply with an optimal design of the production program. When maintaining this requirement difficulties must be expected whenever the joint product with a relatively low selling price and small maximum sales quantity handicaps profit planning because another joint product of the same splitting process with a comparatively very high selling price cannot be produced until its maximum sales quantity is reached. Corresponding considerations are possible when the joint products cannot only be sold but also processed and when the necessary processing quantities of the following production stages are included in the considerations.

Because of these doubts with regard to the methodology one must start from the fact that for an optimal design of a joint production program, the requirement of the standard approach not to produce more of each category of product than can actually be sold or processed, must be given up. Therefore some quantities of output have to be included which can neither be used within the production process nor marketed. These quantities are called surplus or disposal quantities. Because these quantities have to be allowed for within the planning approach entails additional factors which influence an optimal design of the joint production program. These factors include the cost for disposing of the disposal quantities as well as the maximum disposal quantities which set an upper limit on the production of tolerable surplus quantities.

The following investigation is particularly concerned with the question as to what extent the optimal design of a joint production program is affected by variable disposal costs and maximum disposal quantities, while all the other data remain constant.

For this purpose we introduce a generalized formulation approach with parametrically variable disposal costs and maximum disposal quantities. This approach may be regarded as a synthesis of most planning approaches in joint production which can be found in the literature, and contains the latter as special cases. By means of a numerical example we shall demonstrate what changes in profits will possibly have to be put up with, depending on which special case is chosen for the formulation of the problem, and in what way the surplus quantities of individual joint products will then behave. The results obtained are directly applicable to a practical planning situation, since they show that deliberately allowing for surplus quantities is still more profitable, even if comparatively high disposal costs have to be considered which would ordinarily suggest a general avoidance of surplus quantities. In the next section we shall point out how the different model approaches to be found in the literature on joint production take into account the treatment of surplus or disposal quantities and what consequences this might entail.

II. The Treatment of the Surplus or Disposal Quantity Problem in the Literature

The different degree of consideration and the varying treatment of surplus or disposal quantities in the literature approaches to joint production are characterized by the individual method applied and the planning goal pursued. In his cost studies on joint production, v. STACKELBERG assumes that the joint products of a single-stage splitting process with rigid proportions can be bundled into product packages. Under these conditions he proves that the findings from the theory of simple production are also directly applicable to the theory of joint production for the purposes of profit and cost determination as well as profit planning (v. STACKELBERG 1932, p. 57 and p. 60). In this connection disposal quantities are dealt with only in so far as they have to be taken into account for the waste products of a package. Surplus quantities, however, which might additionally occur in the other joint products of a package and may represent main or by-products of the splitting process are neglected. This implies that with respect to the marketable joint products of a package no sales constraints have to be observed. Due to this implication and the fact that v. STACKELBERG confines himself to single-stage splitting processes, he neglects in his approach that at the appearance of surplus quantities and their disposal, the profit and cost rates of a package will change in those cases when the different joint products of a bundle exceed their individual maximum sales quantities or processing quantities. Thus the applicability of the findings from the theory of simple production to problems of program planning in joint production is considerably modified.

In his remarks on program planning with respect to joint production, RIEBEL also assumes that all marketable goods which are manufactured with rigid proportions in the course of a planning period can be sold. He extends v. STACKELBERG's idea of making packages to the cases of multi-stage joint production processes and tries simultaneously to make it accessible to cost covering margin accounting. Therefore, also surplus or disposal quantities will occur only with respect to waste products within one package. The proportional disposal costs connected herewith are reflected in the cost covering margin of a product bundle in a similar form as is the case in v. STACKELBERG's approach. However, because of the multistage production process possibly cost covering margins for product bundles within product bundles will have to be calculated in addition. Surplus quantities of those products are ignored which cannot be immediately regarded as waste due to their very nature. For one thing, it may be possible to refine all quantities of the joint products, except for those of the waste products, by attaching appropriate separate processing stages in such a way that they can all be marketed as the demand is. Another alternative consists of largely making up for the discrepancies which appear between the obtained quantities of the various joint products, by a clever mix or combined processing of the by-products into marketable final products. For both arguments there are important indications in the production of milk and cattle processing, as well as in the fields of crude oil and chemistry (BIETHAHN 1974; MANNE 1958). Yet, it will not be generally feasible to assume that surplus quantities of otherwise usable joint products can always be avoided in one way or another in program planning and are therefore negligible. There are examples demonstrating the opposite. If, however, such surplus quantities and their disposal are to be taken into account from the start, then the application of the cost covering margin accounting to multi-stage joint production within the framework of the package concept becomes almost impossible. In the case of quantity variations the profit rates of the product bundles will change discretely. RIEBEL himself already implies this problem when pointing out that in bottleneck situations, in order to avoid false decisions in program planning due to negative total cost covering margins which might occur, the profit impacts of partial programs of various processing possibilities must always be set against one another (RIEBEL 1971, pp. 757–759).

Compared to the time-consuming and complicated concept of package making, surplus quantities – no matter for what kind of joint production – can formally be dealt with in a much simpler way in the solution approaches of program planning based on process oriented calculation. Here these quantities can be integrated into the balance and quantity continuity conditions as independent variables, and – valued at their disposal costs – they will figure immediately as profit reducing components in the profit function. Nevertheless, the degree of attention paid to surplus or disposal quantities is quite different. MÜLLER-MERBACH (1973, pp. 168–171) for example includes no surplus quantities at all in formulating an optimal production program in multi-stage joint production. Consequently, the production quantities must not exceed the processing quantities and maximum sales quantities of the joint products. This neglect of surplus quantities which in the economic sense corresponds to the unfeasibility of disposal quantities implies that multi-stage joint production is limited merely by the comparatively lowest maximum sales quantity of the goods manufactured. This restriction is propagated retrogradely in all processing quantities of the preceding stages. This simultaneously obstructs the manufacture of joint products which, at comparatively high selling prices, have not yet reached their maximum sales quantity. The sales returns of these would still show a positive profit contribution after substracting the disposal costs for the surplus quantities of other joint products. Unsatisfactorily, the question to what extent the inclusion of surplus quantities has a profit increasing effect is thus left unanswered.

BIETHAHN (1974, p. B167) applies the process oriented approach in a slightly modified form to a joint production problem in the meat industry. It is assumed in the various models that a fixed or a variable sales program with upper and lower limits for the sales quantities is to be optimized with respect to profits or costs. Joint products occur in the butchering processes of the meat and sausage production. In the sense of a maximum usage of material, those by-products which are not suited for selling can subsequently be processed almost entirely into certain kinds of sausage, or they can be refined and then be sold. Within the framework of the simultaneous planning of the purchasing program it is possible to buy more of all quantities of raw products which require additional processing. These unlimited possibilities to buy more, as well as the fortunate circumstance that on account of the manufacturing structures in this branch of production the principle of maximum usage of material prevails to a very high degree, imply that only little importance is attributed to the surplus or disposal quantities of the joint products due to their low production and disposal costs.

Frequently, these quantities can even be further reduced by special offers. Correspondingly, BIETHAHN regards the surplus quantities only as slack variables of the cleavage products and recipes in the quantity balances and quantity continuity conditions of the program approach. They do not occur, however, in the objective function in which the difference between sales and cost price is maximized. No objection is raised to this procedure as long as the surplus quantities, due to their size and their low disposal costs, do not too essentially influence the content of the production program and the corresponding profit. In most cases, however, one will have to reckon with disposal amounts and costs.

KRUSCHWITZ and KILGER were the first to show how the process oriented calculation in its original version is also applicable to program planning in joint production. In their examples they either fix a priori those joint products for which the disposal quantities are to be taken into account (KRUSCHWITZ 1974, p. 104; KILGER 1973, p. 361), or they make the inclusion of the latter dependent on the obvious insignificance of by-products (KILGER 1973, p. 367 and p. 371) which are obtained within the framework of joint processes. While KILGER points out simultaneously that possibly for these disposal quantities upper limits will have to be defined, KRUSCHWITZ explicitly limits the former by inplant capacity bottlenecks. Considering surplus quantities ex ante with respect to only certain joint products is implicitly justified by the fact that for the rest of the by-products no surplus quantities are feasible, the special problem formulation and approach resulting from this are acceptable. Otherwise, it is possible that such an anticipated assessment of the planning situation might entail false decisions with respect to the optimal content of the program. For, as a rule, it is only after having designed a general program planning approach containing surplus variables for all joint products that it will become apparent for what kinds of product disposal quantities must be contended with for the maximum profit. The appearance of disposal quantities with respect to the various by-products can change very quickly by merely varying the level of the unit costs of disposal. In the following the appearance of surplus quantities in joint production and their behavior in view of a uniform variation of the once fixed unit costs of disposal as well as variable maximum disposal quantities will be analysed more closely on the basis of a generalized formulation approach of the process oriented calculation. This will clarify the interdependencies in joint production planning.

III. A General Approach to Planning of Joint Production with Variable Disposal Costs and Maximum Disposal Quantities

1. Formulation of the Approach

The following statements are based on a multi-stage product structure which is assumed to be characterized by multiple, linear joint production with rigid proportions and combined processing. The linearity of the manufacturing structure implies that it does not contain any loops. Because of rigid proportions, problems with respect to the choice of procedure between alternative splitting processes will not be considered. Moreover, we assume that within the framework of production program planning there are no external sources of supply and no storage between the production stages.

Suppose that the production does not entail any fixed costs. Therefore the profit consists solely of the difference between the sales returns and the variable manufacturing cost, as well as the relatively proportional disposal costs. Each product of the various production stages is assumed either to result from processed initial products or from joint processes of the preliminary stage. To simplify matters let the goods of the first kind of product at each stage of production run through the first production quantity indices each. The residual indices are reserved for the goods of the second kind of product. The distinction between these two kinds of products – originating either by combined processing or from joint processes – is necessary, since the capacity constraints which must possibly be allowed for in case of converging or diverging production, and the manufacturing cost for goods resulting from processing or for the inputs of the splitting processes are to be treated qualitatively in a different way. While in the first case they will usually be related to the quantities manufactured, they must be formulated for the inputs in the second case, since the imputed costs for joint products cannot be indicated exactly. Moreover, let us point out here that certain joint products may well originate from different splitting processes of the preliminary stage as well. The product variables of the different stages of production are defined in such a way as to imply the raw materials also for the zero-th stage. Each kind of product has four usabilities at the most: it can be sold, processed, used as input in a joint process, or be disposed of. The actual usage is subject to the conditions of production technology. The sum of these partial quantities – except for the raw materials – must always be equal to the quantity of products manufactured. On the zero-th stage of production no sales and no disposal of raw materials is to be feasible. Thus at this step there are also no revenues and no disposal costs. On the last stage only marketable products and disposal quantities are to be placed. Thus in this stage there will be no manufacturing cost for inputs of joint processes. Marketable intermediate, by- and final products may occur on the stage lying in between. Production bottlenecks will usually result from capacity constraints for labour and working funds and may have to be considered when formulating the problem. The disposal of surplus quantities, however, assumed to be performed outside the plant, does not use one's own production capacities and incurs no cost. As a rule, upper limits have to be observed for the quantities of raw materials used, as well as the sales quantities and disposal quantities of the product.

With these assumptions let us denote by

s = $0, \ldots, l$ the stages of production;

j_s = $1, \ldots, n_s$ the products of stage s which result for $j_s = 1, \ldots, \bar{n}_s$ from combined processing or for $j_s = \bar{n}_s + 1, \ldots, n_s$ from joint processes, and imply the raw materials for $s = 0$;

i_{s+1} = $1, \ldots, m_{s+1}, s = 0, \ldots, l-1$, labour and machinery funds available for joint processes on stage $s+1$;

q_s = $1, \ldots, r_s, s = 0, \ldots, l-1$, labour and machinery funds available for joint processes on stage s;

$x^s_{j_s}$ the quantities of the product j_s which are available or have been produced on stage s;

$x^{sA}_{j_s}, x^{sW}_{j_s}, x^{sE}_{j_s}, x^{sV}_{j_s}$ the partial quantities of $x^s_{j_s}$ which are sold, processed, used in joint processes, or disposed of;

$\bar{x}^{sA}_{j_s}, \bar{x}^{sV}_{j_s}$, $s = 1, \ldots, l$, maximum sales quantities or disposal quantities;

$\bar{x}^0_{j_0}$ the maximum quantities of raw materials used;

$a^s_{j_s j_{s+1}}$, $s = 0, \ldots, l-1$, $j_{s+1} = 1, \ldots, \bar{n}_{s+1}$, the production coefficients with respect to the materials used in processing;

$c^s_{i_{s+1} j_{s+1}}$, $s = 0, \ldots, l-1$, $j_{s+1} = 1, \ldots, \bar{n}_{s+1}$, the coefficients of the use of machinery funds and labour by processing;

$g^s_{i_{s+1}}$ the processing capacities;

$b^s_{j_{s+1} j_s}$, $s = 0, \ldots, l-1$, $j_{s+1} = \bar{n}_{s+1}+1, \ldots, n_{s+1}$, the yield coefficients for joint processes,

$d^s_{q_s j_s}$, $s = 0, \ldots, l-1$, the coefficients of the use of machinery funds and labour by inputs of the joint processes;

$h^s_{q_s}$ the capacities of joint production;

$p^s_{j_s}$, $s = 1, \ldots, l$, the selling prices;

$k^s_{j_s}$, $s = 1, \ldots, l$, the unit costs of processing;

$k^{sE}_{j_s}$, $s = 0, \ldots, l-1$, the unit costs arising by the use of $x^{sE}_{j_s}$ in a splitting process;

$k^{sV}_{j_s}$, $s = 1, \ldots, l$, the unit costs for the disposal of surplus quantities;

Q the profit;

λ, u nonnegative, real-valued parameters for varying the level of the unit costs of disposal and the maximum disposal quantities.

Being based on the process oriented calculation the generalized planning model for a profit-maximal design of the production program in joint production can then be formulated by the following linear approach:

$$\text{(1)} \quad \max Q = \sum_{s=1}^{l} \sum_{j_s=1}^{n_s} p^s_{j_s} x^{sA}_{j_s} - \sum_{s=1}^{l} \sum_{j_s=1}^{\bar{n}_s} k^s_{j_s} x^s_{j_s} - \sum_{s=0}^{l-1} \sum_{j_s=1}^{n_s} k^{sE}_{j_s} x^{sE}_{j_s} - \lambda \cdot \sum_{s=1}^{l} \sum_{j_s=1}^{n_s} k^{sV}_{j_s} x^{sV}_{j_s}$$

under the constraints:

$$\text{(2)} \quad x^{sA}_{j_s} + x^{sW}_{j_s} + x^{sE}_{j_s} + x^{sV}_{j_s} = x^s_{j_s}, \quad j_s = 1, \ldots, n_s, \quad s = 0, \ldots, l;$$

$$\text{(3)} \quad \sum_{j_{s+1}=1}^{\bar{n}_{s+1}} a^s_{j_s j_{s+1}} x^{s+1}_{j_{s+1}} = x^{sW}_{j_s}, \quad j_s = 1, \ldots, n_s, \quad s = 0, \ldots, l-1;$$

$$\text{(4)} \quad \sum_{j_{s+1}=1}^{\bar{n}_{s+1}} c^s_{i_{s+1} j_{s+1}} x^{s+1}_{j_{s+1}} \leq g_{i_{s+1}}, \quad i_{s+1} = 1, \ldots, m_{s+1}, \quad s = 0, \ldots, l-1;$$

$$(5)\quad x_{j_{s+1}}^{s+1} = \sum_{j_s=1}^{n_s} b_{j_{s+1}j_s}^{s} x_{j_s}^{sE} , \quad j_{s+1} = \bar{n}_{s+1}+1, \ldots, n_{s+1} , \quad s = 0, \ldots, l-1 ;$$

$$(6)\quad \sum_{j_s=1}^{n_s} d_{q_s j_s}^{s} x_{j_s}^{sE} \leq h_{q_s} , \quad q_s = 1, \ldots, r_s , \quad s = 0, \ldots, l-1 ;$$

$$(7)\quad x_{j_0}^{0} \leq \bar{x}_{j_0}^{0} , \quad j_0 = 1, \ldots, n_0 ;$$

$$(8)\quad x_{j_0}^{0A}, x_{j_0}^{0V} = 0 , \quad j_0 = 1, \ldots, n_0 ;$$

$$(9)\quad x_{j_l}^{lW}, x_{j_l}^{lE} = 0 , \quad j_l = 1, \ldots, n_l ;$$

$$(10)\quad x_{j_s}^{sA} \leq \bar{x}_{j_s}^{sA} , \quad j_s = 1, \ldots, n_s , \quad s = 1, \ldots, l ;$$

$$(11)\quad x_{j_s}^{sV} \leq u \cdot \bar{x}_{j_s}^{sV} , \quad j_s = 1, \ldots, n_s , \quad s = 1, \ldots, l ;$$

$$(12)\quad x_{j_s}^{s}, x_{j_s}^{sA}, x_{j_s}^{sW}, x_{j_s}^{sE}, x_{j_s}^{sV} \geq 0 , \quad j_s = 1, \ldots, n_s , \quad s = 0, \ldots, l ;$$

with parameters $\lambda, u \geq 0$.

The profit function (1) is composed of the revenues – first term –, the variable manufacturing cost arising by processing and joint processes – second and third term –, and the relatively proportional disposal costs the levels of which are variable – last term –. The quantity balance conditions (2) express that the sum of the partial quantities of a product which have been used in the various ways must be equal to its total quantity available or produced. The quantity continuity conditions (3) ensure that on each stage as much of the various products is available for processing as is required for manufacturing the product quantities of the following stage which originate by that kind of manufacture. Analogously, the yield constraints (5) reflect the relative proportions between the joint processes. (4) and (6) represent the capacity conditions for processing and joint production. The other additional conditions (7) to (12) are self-evident in view of the detailed premises stated before. By extending the problem formulation by the parameters λ and u in the objective function (1) and the constraints (11) it is possible to study somewhat more closely – under otherwise constant conditions – the dependence of the profit-maximal production program in joint production and the occurring surplus quantities upon the levels of the unit costs of disposal and the maximum disposal quantities, and to compare the results of different special cases of the formulation approach with each other.

2. Special Cases of the Approach and Their Economic Importance

Of the boundless possibilities to vary the parameters λ and u in the planning approach (1) to (12) essentially such combinations (λ, u) are interesting which lead to the special cases of program planning in joint production which are already known from the literature – as they have been discussed in Section II –, or else suggest changes in the appearance of disposal quantities with respect to the various joint products. Combinations of this kind to which also the following representation of a numerical example will be confined, have been compiled in Table 9.1; here Λ and U are to be sufficiently large numbers ensuring that at this level of the disposal costs no surplus quantities will appear any more, or the

Table 9.1. Special cases of the planning approach (1) – (12)

λ \ u	0	1	U
0	1 a) Müller-Merbach Kilger	3 a)	2 a) Biethahn
1	1 b) –	3 b) Kruschwitz	2 b) Kilger (v. Stackelberg) (Riebel)
$1 < \lambda < \Lambda$	1 c) –	3 c) analogous to 2 c)	2 c)
Λ	1 d) –	3 d) analogous to 1 a)	2 d) analogous to 1 a)

maximum disposal quantities are no longer binding. Because of the methodically different procedure in the approaches by v. Stackelberg and Riebel these names have been put in brackets in Table 9.1.

In case 1 a) no surplus quantities are feasible and simultaneously, the disposal costs can be assumed to be equal to zero. This case may also be interpreted in such a way that disposal costs are not being allowed for in program planning. A more detailed consideration of the combinations 1 b) – 1 d) for which the level of unit costs of disposal is equal to one or rises constantly for unfeasible surplus quantities, is unprofitable. It would lead to the same result as in case 1 a). The cases 2 a) – 2 d) in which, irrespective of the individual level of the disposal costs, maximum surplus quantities are not relevant to decisions, dominate the cases 3 a) – 3 d) with respect to profit planning. In these cases the additional conditions for the actual maximum disposal quantities may have a restrictive effect. Cases 3 a) and 2 a) differ from cases 3 b) and 2 b) by the fact that in the former two cases the disposal of surplus quantities can be performed free of charge, whereas in the latter two cases disposal costs have to be taken into account. In cases 3 c) and 2 c) the level of the disposal costs can be raised successively in discrete steps, in order to analyse somewhat more closely the further behavior of surplus quantities of the different joint products. Finally in cases 3 d) and 2 d) no more surplus quantities will occur due to prohibitive disposal costs. Thus, from the economic viewpoint, the latter two cases approach to case 1 a) with respect to the result. Here it is particularly interesting to know the minimum level Λ of the disposal costs which under profit aspects justifies the use of the two alternative planning concepts, i.e. for one thing disposal quantities are formally included, for another thing, however, they remain ex ante unconsidered. With respect to cases 3 c) and 2 c) we can confine ourselves to the subcases of 2 c), since from a certain level of disposal costs the maximum surplus quantities in 3 c), too, will be no longer relevant to decisions, and then there will be an analogous course of phenomena. Cases 1 a) and 2 a) represent borderline cases of profit-maximal program planning in joint production. Case 1 a) will regularly imply the lowest profit, since the infeasibility or neglect of surplus quantities entails the greatest reduction of the decision space. For 2a), however, always the highest profit may be expected, since the surplus quantities are practically unbounded and there are no disposal costs. The surplus quantities, however, do not exceed the inevitable bounds, because every production entails costs thereby reducing profit. The rest of the cases in Table 9.1 lie between these two borderline cases. The above considerations are to be illustrated by a numerical example in the following Section IV.

IV. Numerical Example

1. Production Pattern, Data, and Problem Formulation

Let the production pattern underlying the numerical example be characterized by Fig. 9.1 according to the statements of Section III.1. The input-output relationships between the products are represented by arrows, the solid ones standing for diverging splitting processes and the dashed ones for converging processes. For the sake of simplicity and clarity the appearing products have not additionally been marked by an index of the stage of production. They have simply been numbered consecutively. Product 1 is the only raw material used in production of which the final products 8 to 13 are manufactured by the intermediate products 2–7 on the subsequent stages of production. The intermediate products 2 to 7 are to be used only as inputs for joint processes or else for processing, but they are assumed not to be marketable. Capacity constraints of processing for labour

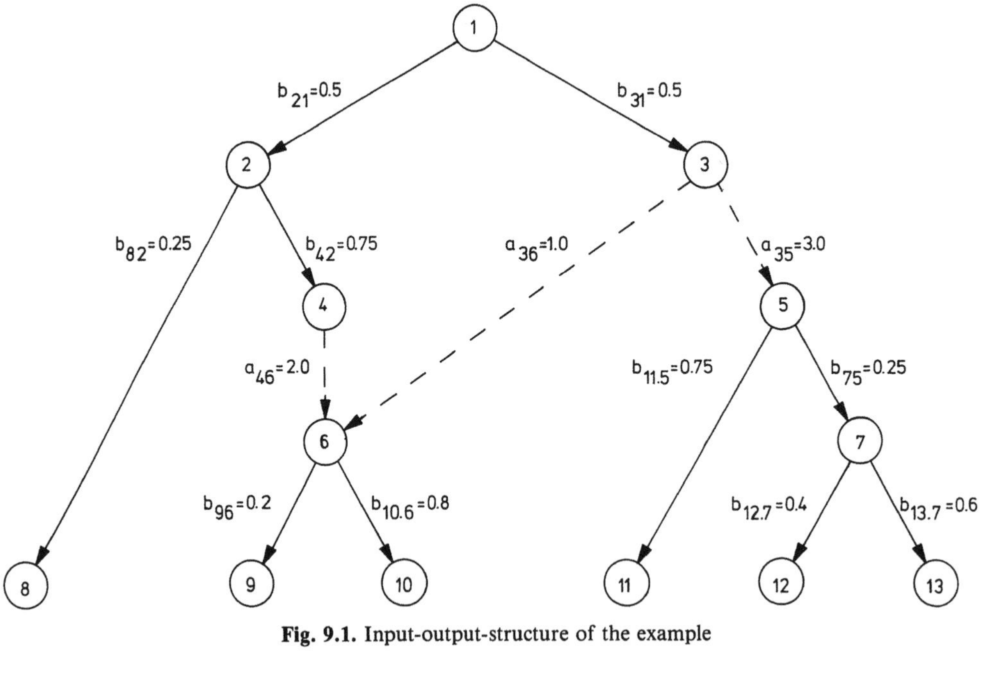

Fig. 9.1. Input-output-structure of the example

and working funds are neglected, since they are supposed to have no bottleneck effects on production. Let the similar hold also for the use of resources in the joint processes. Only with respect to the quantities of the intermediate products 6 and 7 used in splitting processes we assume that their processing is performed in common by the same aggregate and is subject to a corresponding capacity constraint. Production and sales data of the numerical example have been compiled in Table 9.2 using the symbols introduced in Section III.1; here the quantities are calculated in units, and selling prices as well as manufacturing and disposal costs in DM/unit. The unit costs of processing are listed in the columns of those products which originate by processing. The unit costs of the joint processes, however, are specified together with those products whose partial quantities are suitable for inputs of splitting processes. From the last line of Table 9.2 the capacity limit of the aggregate can be seen in which the inputs of the products 6 and 7 are processed in common; the corresponding coefficients of capacity use are listed in line d_j. As far as the intermediate products 2 to 7 are concerned, half the unit costs of production of the respective production processes – splitting or processing – from which these products originate have been rated as unit costs of disposal. For the final products 8 to 13 it has been assumed that their unit costs of disposal total half the sales prices. This choice will later facilitate the comparability of the results which come from different variations of the level of the unit costs of disposal. The production and yield coefficients are entered at the arrows of Figure 9.1.

Now, if a numbering of the objective function and the additional conditions is chosen which is analogous to the general planning model, then the program planning approach for the numerical example reads as follows:

$$(1') \quad \begin{aligned} \max Q = {} & 50x_8^A + 17x_9^A + 100x_{10}^A + 26x_{11}^A + 50x_{12}^A + 190x_{13}^A \\ & -(4.5x_5 + 2.5x_6) - (2x_1^E + 1x_2^E + 4x_5^E + 3x_6^E + 5x_7^E) \\ & -\lambda(1x_2^V + 1x_3^V + 0.5x_4^V + 2.25x_5^V + 1.25x_6^V + 2x_7^V \\ & + 25x_8^V + 8.5x_9^V + 50x_{10}^V + 13x_{11}^V + 25x_{12}^V + 95x_{13}^V) \end{aligned}$$

under the constraints

$$(2') \quad x_j^A + x_j^W + x_j^E + x_j^V = x_j \;, \quad j = 1, \ldots, 13 \;;$$

$$(3') \quad 3x_5 + 1x_6 = x_3^W \;, \quad 2x_6 = x_4^W \;;$$

(4') not applicable;

$$(5') \quad \begin{cases} x_2 = 0.5\,x_1^E \;, & x_3 = 0.5x_1^E \;, & x_4 = 0.75x_2^E \;, & x_7 = 0.25x_5^E \;, \\ x_8 = 0.25x_2^E \;, & x_9 = 0.2x_6^E \;, & x_{10} = 0.8x_6^E \;, & x_{11} = 0.75x_5^E \;, \\ x_{12} = 0.4\,x_7^E \;, & x_{13} = 0.6x_7^E \;; \end{cases}$$

$$(6') \quad 2x_6^E + 4x_7^E \leq 3000 \;;$$

$$(7') \quad x_1 \leq 10000 \;;$$

$$(8') \quad x_j^A = 0 \;, \quad j = 1, \ldots, 7 \;; \quad x_1^V = 0 \;;$$

Table 9.2. Data for the numerical example

Product j	1	2	3	4	5	6	7	8	9	10	11	12	13
Maximum sales volumes $\bar{x}_j^A$	–	–	–	–	–	–	–	700	200	900	500	50	150
Maximum quantities of disposal $\bar{x}_j^V$	0	350	100	150	120	200	60	40	30	90	50	50	100
Maximum quantities of raw material $\bar{x}_j$	10000	–	–	–	–	–	–	–	–	–	–	–	–
Use of labour and working funds by joint processes d_j	–	–	–	–	–	2	4	–	–	–	–	–	–
Selling prices p_j	–	–	–	–	–	–	–	50	17	100	26	50	190
Unit costs of processing k_j	–	–	–	–	4.5	2.5	–	–	–	–	–	–	–
Unit costs of joint processes k_j^E	2	1	–	–	4	3	5	–	–	–	–	–	–
Unit costs of disposal k_j^V	–	1	1	0.5	2.25	1.25	2	25	8.5	50	13	25	95

Capacity of joint production for the inputs 6 and 7: $h = 3000$.

(9′) $x_j^W = 0$, $j = 1, 2, 5, 6, \ldots, 13$; $x_j^E = 0$, $j = 3, 4, 8, \ldots, 13$;

(10′) $x_8^A \leq 700$, $x_9^A \leq 200$, $x_{10}^A \leq 900$, $x_{11}^A \leq 500$, $x_{12}^A \leq 50$, $x_{13}^A \leq 150$;

$$(11') \quad \begin{cases} x_1^V \leq u \cdot 0 , & x_2^V \leq u \cdot 350 , \quad x_3^V \leq u \cdot 100 , \quad x_4^V \leq u \cdot 150 , \quad x_5^V \leq u \cdot 120 , \\ x_6^V \leq u \cdot 200 , & x_7^V \leq u \cdot 60 , \quad x_8^V \leq u \cdot 40 , \quad x_9^V \leq u \cdot 30 , \quad x_{10}^V \leq u \cdot 90 , \\ x_{11}^V \leq u \cdot 50 , & x_{12}^V \leq u \cdot 50 , \quad x_{13}^V \leq u \cdot 100 ; \end{cases}$$

(12′) $x_j, x_j^A, x_j^W, x_j^E, x_j^V \geq 0$, $j = 1, \ldots, 13$;

with parameters $\lambda, u \geq 0$.

The differences in the formulations of the additional conditions (8′) and (9′) as against (8) and (9) in Section III.1 are due to the assumptions underlying the numerical example that the intermediate products 2 and 7 cannot be marketed, and only 3 and 4 are designated for processes of combined processing, but are not to be used as inputs for joint processes.

2. Results of Calculations

For the calculations it turned out beforehand that with respect to the unit costs of disposal and maximum disposal quantities given in Table 9.2 the levels $\lambda = \Lambda = 375$ or $u = U = 2$ suffice that no surplus quantities are produced within the production program or that the vector of the upper limits for the surplus quantities is no longer relevant to the decisions. In the subcases of 2c), however, the λ-values 4.5, 5, 20, 80, and 374 have proved especially remarkable with respect to the changes of the results, but with this not the entire interesting variation spectrum for this parameter of the disposal cost level is to be exhausted altogether at the same time. The calculated results for the numerical examples have been arranged according to the case differentiations of Table 9.1 and compiled in Table 9.3. In order to save space the computer code has been kept up for the product variables, their different uses and indexing; the code can be read accordingly when taking the first letter after X as upper index and the following number as subindex. On the basis of Table 9.3 the calculated results may thus be commented on or economically interpreted as follows.

The difference in profits between case 1a) in which no surplus quantities are feasible or neglected ex ante in planning, and the profit maximal approach to production program planning for case 2a) in which no maximum disposal quantities have to be taken into account and the surpluses obtained can be disposed of at no cost, amounts to DM 25,540 or about 23% related to the more unfavourable one of the two profit results. As against the deviation in case 3b) which starts from binding maximum disposal quantities and defined unit costs of disposal – and this may reasonably be assumed to be the standard case in practice – the deviation in planning approach 1a) totals after all as much as DM 22,206 or about 20% with respect to the profit, if in program planning in joint production the conceptional error is committed not to take surplus quantities in account. Due to complete suppression of surplus quantities in case 1a) only product 12, which is quite cheap as compared with product 13 and results from the

Table 9.3. Calculated results for the numerical example

	1a) $u=0$ $\lambda=0$	3a) $u=1$ $\lambda=0$	3b) $u=1$ $\lambda=1$	2a) $u=U=2$ $\lambda=0$	2b) $u=U=2$ $\lambda=1$	2c) $u=U=2$ $\lambda=4.5$	$\lambda=5$	$\lambda=20$	$\lambda=80$ ($\lambda=374$)	2d) $u=U=2$ $\lambda=\Lambda=375$
Q	109735	134145	131941	135275	132708	126242	125971	122633	117914 (109780)	109735
$X1$	4800	6620	6220	6750	6250	6100	5600	5600	5333.3	4800
$X2$	2400	3310	3110	3375	3125	3050	2800	2800	2666.7	2400
$X3$	2400	3310	3110	3375	3125	3050	2800	2800	2666.7	2400
$X4$	1800	2200	2220	2250	2250	2100	2100	2100	2000	1800
$X5$	500	733.3	666.7	750	666.7	666.7	583.3	583.3	555.5	500
$X6$	900	1110	1110	1125	1125	1050	1050	1050	1000	900
$X7$	125	183.3	166.7	187.5	166.7	166.7	145.8	145.8	138.8	125
$X8$	600	740	740	750	750	700	700	700	666.7	600
$X9$	180	222	222	225	225	210	210	210	200	180
$X10$	720	888	888	900	900	840	840	840	800	720
$X11$	375	550	500	562.5	500	500	437.5	437.5	416.7	375
$X12$	.50	73.3	66.7	75	66.7	66.7	58.3	50	50	50
$X13$	75	110	100	112.5	100	100	87.5	75	75	75
$XA8$	600	700	700	700	700	700	700	700	666.7	600
$XA9$	180	200	200	200	200	200	200	200	200	180
$XA10$	720	888	888	900	900	840	840	840	800	720
$XA11$	375	500	500	500	500	500	437.5	437.5	416.7	375
$XA12$	50	50	50	50	50	50	50	50	50	50
$XA13$	75	110	100	112.5	100	100	87.5	75	75	75

*XW*3	2400	3310	3110	3375	3125	3050	2800	2800	2666.6	2400
*XW*4	1800	2220	2220	2250	2250	2100	2100	2100	2000	1800
*XE*1	4800	6620	6620	6750	6250	6100	5600	5600	5333.3	4800
*XE*2	2400	2960	2960	3000	3000	2800	2800	2800	2666.7	2400
*XE*5	500	733.3	666.7	750	666.7	666.7	583.3	583.3	555.5	500
*XE*6	900	1110	1110	1125	1125	1050	1050	1050	1000	900
*XE*7	125	183.3	166.6	187.5	166.7	166.7	145.8	125	125	125
*XV*2	0	350	150	375	125	250	0	0	0	0
*XV*3	0	0	0	0	0	0	0	0	0	0
*XV*4	0	0	0	0	0	0	0	0	0	0
*XV*5	0	0	0	0	0	0	0	0	0	0
*XV*6	0	0	0	0	0	0	0	0	0	0
*XV*7	0	0	0	0	0	0	0	20.8	13.8	0
*XV*8	0	40	40	50	50	0	0	0	0	0
*XV*9	0	22	22	25	25	10	10	10	0	0
*XV*10	0	0	0	0	0	0	0	0	0	0
*XV*11	0	50	0	62.5	0	0	0	0	0	0
*XV*12	0	23.3	16.7	25	16.7	16.7	8.3	0	0	0
*XV*13	0	0	0	0	0	0	0	0	0	0

same joint process, reaches its relatively low maximum sales quantity, thus restricting all other production and sales quantities of the program. Understandably, case 2 a) shows the highest use of resources x_1^E which, however, in case of restrictive maximum surplus quantities – case 3 a) and 1 a) – or rising level of unit costs of disposal – cases 2 b) to 2 d) – will naturally decrease together the total profit. The capacity constraint of the joint processes, which is valid for the processing of the input quantities of the intermediate products 6 and 7 that are used in splitting processes, proves to be binding only in case 2 a), becoming irrelevant to decisions in the rest of the cases due to the maximum disposal quantities to be observed or the level of the unit costs of disposal. Yet in case 2 a) all final products, except 13, reach their maximum sales quantities, while the maximum disposal quantities – if they were binding as in case 3 a) under otherwise equal conditions – would be exceeded for the products 2, 8, and 11; keeping to the maximum disposal quantities in case 3 a) implies a loss of profit amounting to DM 1130, but this loss would naturally grow with smaller maximum disposal quantities. Correspondingly, cases 2 b) and 3 b) can be compared with each other, the profits of which differ by an amount of DM 767. Furthermore, comparing cases 2 b) and 3 b) with cases 2 a) and 3 a) shows that the assumed standard level of the unit costs of disposal is sufficient that no more disposal quantities are obtained with respect to product 11, while simultaneously the surplus quantities of the intermediate product 2 are drastically reduced; by this, also the feed of raw material 1 is considerably decreased. For $\lambda = 4.5$, however, in case 2 c) the surplus quantities of product 2 increase again compared with case 2 b), but these are stopped now with respect to final product 8, since their disposal has become comparatively too expensive now; thus the disposal is carried out directly with regard to initial product 2, while simultaneously the total production of 4 decreases.

With the production and sales data taken as a basis the final product 13 is not disposed of in any of the cases considered, since its production quantity always remains below the maximum sales quantity. Compared with this special statement which is subject to the specific data structure it can, however, generally be said that with respect to products resulting from combined processes – as is the case in the example concerning the intermediate products 5 and 6 – no disposal quantities will be obtained and may consequently be neglected also in the program planning approach when such products are concerned. Here it is quite reasonable to apply the following method: The variable costs of processing are charged directly to the quantities of production resulting from processing, which is impossible when joint products are concerned, since the imputed costs cannot be determined. Thus, due to reasons of cost only as many quantities are manufactured by processing as are required for further production or can be marketed. Then the disposal of surpluses must be carried out – if at all – with regard to the initial products; those are the intermediate products 3 and 4 in the example. Understandably, with respect to these products 2 and 8 show disposal quantities, since otherwise one would reduce the inputs of 1 and 2 correspondingly. Thus, on account of cost aspects it may generally be stated that always at least one joint product of each splitting process will possess no surplus quantities. When there are no disposal quantities with respect to 2 and 8, then surpluses of 3 and 4 will be prevented later on account of the level of disposal costs.

For the further subcases of 2c) it is noteworthy that from $\lambda = 5$ on no more surpluses will be obtained for input 2. From $\lambda = 20$ disposal quantities are also stopped with respect to final product 12, but for the first time they appear with regard to intermediate product 7 whose surplus disposal is comparatively less expensive, and simultaneously the sale of product 13 is reduced. If $\lambda = 80$ the surplus quantities of 9 are equal to zero; since according to the above considerations no disposal quantities will be obtained for product 6, also the production volume of final product 10 is reduced. In case 2c) one just obtains the same production and sales quantities for $\lambda = 374$ as are obtained for $\lambda = 80$, but the profit is reduced by DM 8134 on account of the considerably increased disposal costs. It is only with $\lambda = 375$ that in case 2d) the prohibitive level of disposal costs is reached which allows no more surplus quantities with regard to the intermediate and final products, thus the result corresponds to the one obtained in case 1a). Economically, however, this means that the original unit costs of disposal assumed in the numerical example with practically unlimited disposal quantities must rise to more than 187 times the manufacturing cost or selling prices of the products concerned, in order to be able to accept a formulation approach of the kind of 1a) which does not allow or neglects disposal costs for optimal program planning in joint production. Here it should be remembered that the unit costs of disposal for the standard case ($\lambda = 1$) which had been entered in Table 9.2 are half the unit costs of production or the selling prices.

V. Final Remarks

The findings concerning the allowance for surplus or disposal quantities in optimal program planning in joint production may be used equally for methodological as well as for practical purposes. Stressing the methodological aspects it can be said that the statements and their illustration by a numerical example have shown that the quality of the planning result is dependent upon the way in which the surplus quantities of the joint products are dealt with in various formulation approaches. Under profit aspects, as a rule, very bad planning results will have to be contended with if the integration of disposal quantities with regard to splitting products is formally neglected, or if it is ex ante assumed that no surpluses are obtained for certain joint products. The generalized problem approach with parametrically variable unit costs of disposal and maximum surplus quantities, which has been chosen here, enables to explain the differences in the results emerging from the qualitatively different problem formulations and to compare them with each other. Furthermore, the use of the general planning model with systematic variations of the level of unit costs of disposal in identical step length offers a chance of determining by parametric programming (GAL 1973, p. 134 and p. 211) those levels for which the appearance of disposal quantities of certain joint products is economically no longer acceptable. Possibly then one has to start already disposing of initial products or reducing the inputs. Thus indications are obtained simultaneously with respect to the maximum surplus disposal costs which can just be accepted in order to allow disposal quantities of certain split products.

The last remark refers directly to those points which might be of special interest from the perspective of practice. From calculations performed on the basis of the generalized model it can be seen in individual cases to what extent and from which level of unit costs of disposal it may be worthwhile for a manufacturer using joint processes to build up his own processing or disposal capacities for by- and waste products. In this way he could avoid externally defined maximum disposal quantities and keep his own production program planning as flexible as possible. From this, two special considerations can be derived which already today, because of increasing environmental pollution by pollutants from industrial production, are proving quite important for many chemical manufacturers. For one thing, the question is in what way the design of the profit-maximal production program will be affected when certain surpluses can no longer the disposed of free of charge. What cost rates will then be still acceptable in order to maintain the optimality of the program design once adopted? The other problem is concerned with the consequences which must possibly be drawn with respect to a re-design of the production program if one has to change from one disposal system to another which may be more expensive or limited. The calculated results of the generalized approach provide practical guidelines for both cases. In the individual decision case these guidelines will help the decision maker to develop a sense of the ranges of acceptable unit costs of disposal, i.e. ranges which do not require the feasibility of surpluses to be excluded. An improvement in problem solving should be expected – from both the methodological as well as the practical point of view.

References

Albach, H.: Produktionsplanung auf der Grundlage technischer Verbrauchsfunktionen, in: Brandt, L. (ed.): Arbeitsgemeinschaft für Forschung des Landes Nordrhein-Westfalen, Nr. 105, Köln-Opladen 1962, pp. 45–98.

Biethahn, J.: Die Bestimmung des optimalen und praktikablen Einkaufs- und Produktionsprogramms bei variabler Kuppelproduktion und vorgegebenem Absatzprogramm, in: Zeitschrift für Operations Research, 1974, pp. B167–B183.

Fandel, G.: Zum Stand der betriebswirtschaftlichen Theorie der Produktion, in: Zeitschrift für Betriebswirtschaft, 1980, pp. 86–111.

Gal, T.: Betriebliche Entscheidungsprobleme, Sensitivitätsanalyse und Parametrische Programmierung, Berlin-New York 1973.

Jacob, H.: Produktionsplanung und Kostentheorie, in: Koch, H. (ed.): Zur Theorie der Unternehmung, Festschrift zum 65. Geburtstag von Erich Gutenberg, Wiesbaden 1962, pp. 205–268.

Kilger, W.: Optimale Produktions- und Absatzplanung, Opladen 1973.

Kruschwitz, L.: Zur Programmplanung bei Kuppelproduktion, in: Zeitschrift für betriebswirtschaftliche Forschung, 1974, pp. 96–109.

Manne, A.S.: A linear programming model of the U.S. petroleum refining industry, in: Econometrica, 1958, pp. 67–106.

Müller-Merbach, H.: Operations Research, München 1973.

Riebel, P.: Kuppelprodukte, Kalkulation der, in: Handwörterbuch des Rechnungswesens, Stuttgart 1970, cols. 994–1006.

Riebel, P.: Zur Programmplanung bei Kuppelproduktion, in: Zeitschrift für betriebswirtschaftliche Forschung, 1971, pp. 733–775.

Riebel, P.: Kuppelproduktion, in: Handwörterbuch der Produktionswirtschaft, Stuttgart 1979, cols. 1009–1022.

Stackelberg, H.v.: Grundlagen einer reinen Kostentheorie, Wien 1932.

Part IV

Cutting Stock Management and Trim Loss Optimization in Industry

Chapter 10

Production Theoretic Foundation of Cutting and Related Processes

HARALD DYCKHOFF

I. Introduction

By many industrial operations, objects are cut up into smaller pieces. Usually, different kinds of pieces are unavoidable so that they are joint products. Such cutting (stock) processes especially exist in those branches of industry in which materials are produced or processed which, because of their natural state or by reason of the prevailing production conditions, are to be found in a definite geometric form, but which have to be cut to size to meet customers' requirements or the needs of the firm's own further production stages. Raw materials of this sort are iron, steel, other metals, paper and pulp, glass, wood, textiles, and synthetics. As a result of stagnating development, intensified international competition, and especially of rising raw material prices several of these often material-intensive branches are faced with stiff competition. Rationalization activities which lead to an improvement in the utilization of the raw materials can, therefore, have a decisive bearing on the success of an enterprise.

An improvement in the utilization of the raw materials can be achieved through a decrease in the number of leftover pieces, i.e. by minimizing the trim loss. Problems regarding this or related aims in the field of cutting up material goods are normally known as cutting (stock) or trim loss problems.

Although cutting problems have attracted a relatively large amount of attention from the point of view of production management, operations research, engineering, and computer sciences (cf. the surveys of GOLDEN 1976; HINXMAN 1980 and COFFMAN et al. 1984 as well as Chapter 12 of this book), there has been paid less attention to studies of the production- and cost-theoretical side. Furthermore, cutting processes are also of importance because of the fact that they can represent a more general type of production process, a type which, with regard to the whole area of environmental protection and its effects on business, has been unduly neglected in the production theory of industrial management: elastic joint production. This is because the leftovers, after cutting to size has taken place, form the residue of a joint process, even if they are to be regarded as waste. The elasticity follows from a high degree of flexibility regarding the arrangement of cutting processes. This flexibility rests on the possible choice between different cutting patterns. Indeed, it is a very elastic joint production, even if it is not continuously variable. A characteristic of many cutting problems arises from the breaches of the usual assumption of production theory that the amounts of material can be divided arbitrarily.

The aim of this paper is to develop some basic concepts for a production theory of cutting and related processes. By means of an axiomatic-like approach, i.e. a systematic analysis of a few assumptions placed at the beginning, models can be derived which usually form the starting point in the literature on the subject. In this way the conditions for the application of the models in practice can be better recognized, and the connection between different approaches becomes clearer. Some of the assertions are also of importance for more general processes of elastic joint production.

First, several basic premises and assumptions, in particular concerning the characteristics of the goods and of the production process, will be introduced and discussed (Section II). They define the "technologies of pure cutting processes". The focus here is on questions of efficiency (Section III). For this purpose, the economic principle has to be generalized in such a way as to make allowance for surplusage and waste which arise in joint processes. The most assertive statements concerning the fundamental structure of the relevant production models can be gained for the case of one-dimensional cutting processes. Their production models can be formulated in various versions building the basis for corresponding cost and proceeds models (Section IV). The paper concludes with some comments on extensions and further aspects (Section V).

II. Fundamental Characteristics of the Goods and of the Production Process

The issues under consideration are the quantitative connections between the "large objects", which are cut up, and the "small objects" cut from them. They build the basis for the economic organization of the underlying cutting processes. The objects cut up are input goods (production factors) and the parts cut from them are output goods (products). For the purposes of a production theory of pure cutting processes only the connection between these two groups will be analysed and other factors which might have a bearing – e.g. the production equipment for the cutting procedure, and the workers using this equipment – will be left out of consideration for the present. A cutting process is then, in the main, characterized by the input of large objects and the output of parts from these objects. Fig. 10.1 shows, for example, a simple (two-dimensional) process in which a rectangle (input) is cut up into seven smaller rectangles (output) according to the pattern indicated by the dotted lines.

Here, I will use the term "cutting process" also in cases of non-material "large" and "small" objects, e.g. in cases of packing or loading processes when the space of containers or pallets is "cut up" into parts which then are occupied by (small) objects to be packed or loaded or are "trim loss". (For a more thorough discussion of this topic see DYCKHOFF 1987.) By concentrating on the input/output relations the actual cutting procedure is explicitly ignored and is treated as a black box. One main difficulty, however, consists in finding suitable patterns to cut the output out of the input. In the case of multidimensional cuts this can lead to serious geometrical problems. For example, it is at present not

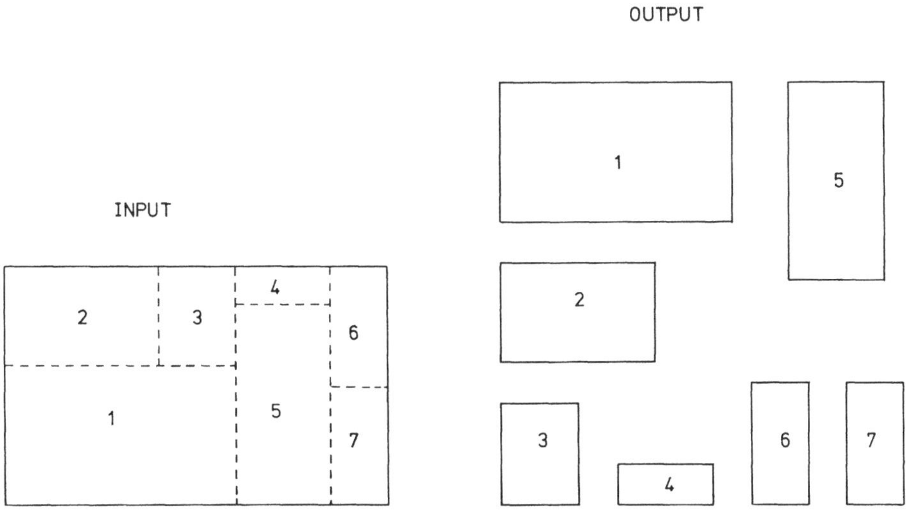

Fig. 10.1. Two-dimensional cutting process

known (cf. GARDNER 1979, 1981) whether it is possible to cut 11 unit squares (lateral length 1) from a single square with an area of 15 (lateral length about 3.873). The geometrical difficulties are, however, often simplified by technological or organizational conditions which have a qualifying effect, for example, the fact that in practice a precise differentiation of the measurements is not possible, or the fact that often only "guillotine cuts" are permitted: these are single straight cuts from one end to another with which an object is divided into (as a rule) two parts.

1. The Characteristics of the Goods

Production theory normally assumes that goods can be divided arbitrarily. "Goods" in this sense are to be understood as things like bulk goods, which can be divided into all sorts of different portions (and can then be joined up again into other larger units) without this manipulation altering their character. This is not only impossible in the case of piece goods such as computers and cars, it is also impossible where things such as lengths of wood or cloth, steel pipes, or sheet metal are concerned, because these are not arbitrarily divisible in the above stated sense. Half a car is no longer a car. Half of a ten-meter long steel pipe is a steel pipe, that is, provided it was cut crossways and not lengthways. Nevertheless, as long as two five-meter pipes cannot be joined to form one ten-meter pipe (at least not without additional costs), pipes of different lengths in principle represent different types of goods. The joining up of shorter pipes to form one longer one, e.g. by welding, is of itself a production process which the theory of production seeks to describe with a definition of the goods and therefore cannot a priori be inseparably mingled with this definition.

It is, therefore, important that arbitrary divisibility of goods cannot be presupposed in its usual meaning. This is a characteristic of cutting processes as production. While it is correct to say that in principle input objects can be divided into smaller pieces, these smaller pieces, as in effect all objects, cannot be joined together to form larger units, at least not without further production processes being called for. Such further production processes are, however, not a subject under discussion here.

Only those goods will be examined which are connected with one another either because they can be divided (cut up) to form one of the other goods, or because they are manufactured by means of the division of one of the other goods. As far as material goods are concerned it is a question of pure shaping procedures by means of which the external form of the goods is altered, but not their material quality. Apart from their geometrical form the goods correspond with one another in all essential qualities, e.g. with textiles, the type of fabric, thickness, colour, and design (as far as these qualities are relevant to decision-making). It is therefore the geometrical definition which is characteristic of these goods:

(G1) Geometrically defined goods

Each of the goods corresponds abstractly to a geometric "figure" in a given, finitely dimensioned space of real numbers. Goods of the same figure are homogeneous. Therefore distinct types of goods are defined solely by means of the figures the goods take, i.e. by their form, orientation and size.

All goods can be represented as geometrical objects in one and the same space which is defined by the dimensions relevant to the division. Hence, the goods are characterized through this representation as to those qualities relevant to decision-making. The space can be one-, two-, or three-dimensional, and in the case of abstract cutting processes can have even more dimensions. It is to be assumed that each geometrical object is connected, i.e. consists of one piece only, and that the objects display the same proportions throughout as the underlying space, i.e. that each figure possesses a mathematical interior whose closure is identical with the figure.

The assumption G1 abstracts from the multitude of goods used in concrete cutting processes through concentration on a few, essential goods. All production factors except for the raw materials and all products apart from the parts as cutting products are ignored. The majority of the models and methods developed for cutting problems use this hypothesis implicitly (or at least the simpler versions do). Without it a first production-theoretical investigation would hardly lead to specific statements of sufficient sharpness and strength.

In accordance with G1 the goods can be sorted into different homogeneous types dependent upon their figure, i.e. their measurements (form, size) and their orientation. Goods of one and the same kind make up geometrical objects with an identical figure in the relevant dimensions. Goods with varying figures belong to different goods-types, e.g. rectangles with varying lengths or widths. As far as one-dimensionality is concerned the types of goods are clearly defined through the lengths of the figures (which are intervals on the real line).

In practice the sizes of the figures in question are always limited upwards and downwards. In the same way the exactness in the gauging and calculating of the shapes is also limited.

(G2) A finite number of goods-types

There exist only a finite number of different types of goods (figures). All figures can be measured – at least approximately – in natural numbers.

Because of the limited accuracy the occurence of the goods-types is usually limited to a controllable amount and rarely reaches the thousands. These often consist of less than a hundred really relevant figures. In practice it is always possible, even if unnecessary, to display all the measurements, which are of interest, as natural numbers by using a suitable unit of measure. This also applies, approximately, to those objects which have an irregular form. In this case the object is slid into the positive orthant of the considered space whose dimensions are relevant to the cutting process. By fixing suitable units of measurement for the co-ordinate axes it is possible to construct a lattice design whose points are defined by the positive integer co-ordinates, and which can be made arbitrarily fine-meshed by adjusting the scale. A figure is then clearly depicted within the framework of the given degree of accuracy by the (discrete and finite number of) lattice points which it contains. Figs. 10.2a and 10.2b illustrate this by each showing the same two-dimensional objects, firstly with a coarse and then with a fine lattice. As far as their discretization is concerned the figures in Fig. 10.2a are identical; in Fig. 10.2b, however, they are different because of the finer differentiation.

Let G indicate the set of the different goods-types (figures) which are met with. If one figure is geometrically contained in another one – which is, of course, larger – it is, in principle, always conceivable that the larger figure can be so divided that the smaller figure appears with others as a result of this division. If this is not possible in a concrete case it is because of additional existing technical or organizational conditions which limit the admissibility of cutting patterns. If there exists a possible pattern for figure g, and figure g contains, among others, figure h, this defines a production connection between g and h in such a way that there is a production process in which g is input and h is output.

(G3) Non-cyclic production connection

There exists a non-cyclic relation $\langle Z \rangle$ to the goods-types so that type h can be the direct result of a possible division of another type, g, if and only if

$$g \langle Z \rangle h$$

is valid. This means that g contains h as a possible result of a division and that h forms a geometrical part of g.

Interpreted in another way "$g \langle Z \rangle h$" expresses that h was made from g, or that g can be used to produce h, or, more precisely: There is a possible production procedure in which goods-type g enters as a factor and goods-type h results as a product, i.e. in which g is input and h output. Herewith production procedure is to be understood as an elementary operation, i.e. not consisting of further

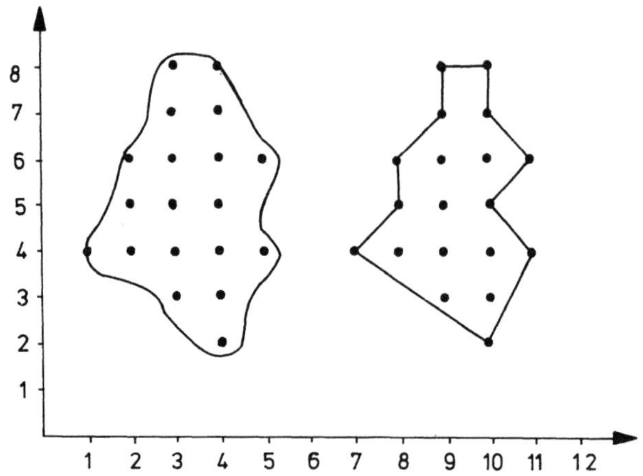

Fig. 10a. Discrete characterization of irregular figures: coarse lattice

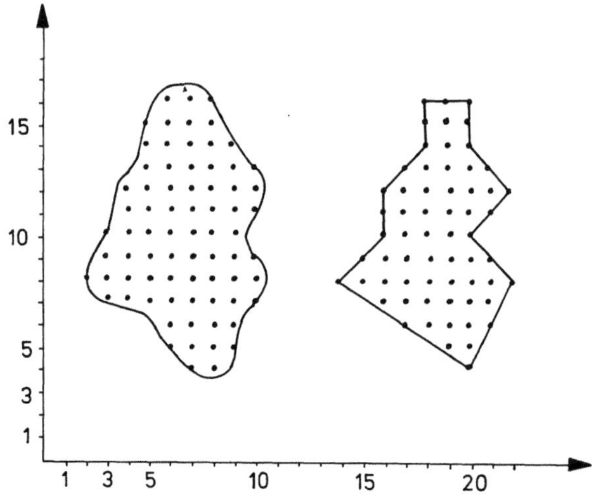

Fig. 10.2b. Discrete characterization of irregular figures: fine lattice

procedures, in which input goods of one or more types are transformed into output goods of one or more types being not necessarily different from the input (see also below).

The postulate of a non-cyclic relation says that for an arbitrary natural number m of (not necessarily different) types of goods $g_1, \dots, g_m$ out of G, "$g_1 \langle Z \rangle g_2$, $g_2 \langle Z \rangle g_3, \dots, g_{m-2} \langle Z \rangle g_{m-1}$, $g_{m-1} \langle Z \rangle g_m$" cannot apply concurrently with "$g_m \langle Z \rangle g_1$". "Non-cyclic" implies that reflexivity is excluded but asymmetry is included: no goods-type can be input and output in one and the same procedure; where there is a production procedure whereby goods-type g is input and goods-type h output there cannot be a procedure whereby h is input and g output. That is to say that goods of a given figure can only be transformed by producing goods of a smaller figure out of them. The interrelation "$g\langle Z\rangle h$" implies that the goods-type h displays a figure which is contained in that of goods-type g; in other words it is a geometrical part of g. (Because of possible technical or other conditions the reverse of the implication does not necessarily need to apply.)

In general, the relation $\langle Z\rangle$ is not transitive. If, for example, a rectangle with a lateral length 2 can be cut from a rectangle with a lateral length 3 by using guillotine cuts with a single blade only, two successive cutting operations are necessary; under these circumstances there is no elementary cutting operation which can do the same.

It is known that a graph can be drawn of every relation. In the case under consideration the different goods-types are the nodes of the graph, whereby an arrow (a directed edge) exists between goods-type g and goods-type h if and only if $g\langle Z\rangle h$ is valid, i.e. when h can be manufactured as an immediate result of a division of g. The graph describes the (direct) connection in production between the types of goods, and will be called a connection graph.

The difference between single-stage and multi-stage connection graphs may be illustrated by means of a two-dimensional example. The input objects all have the same figure, namely a square with a lateral length 3. All rectangles with an integer lateral length of 3 or less are suitable as output, whereby the rectangles have a fixed orientation, i.e. they may not be turned. Hence there are nine possible figures as goods-types: 3×3, 3×2, 2×3, 2×2, 3×1, 1×3, 2×1, 1×2, 1×1 (length and width, respectively). Fig. 10.3a shows the connection graph for a single-stage cutting process in which the square 3×3 is the only input out of

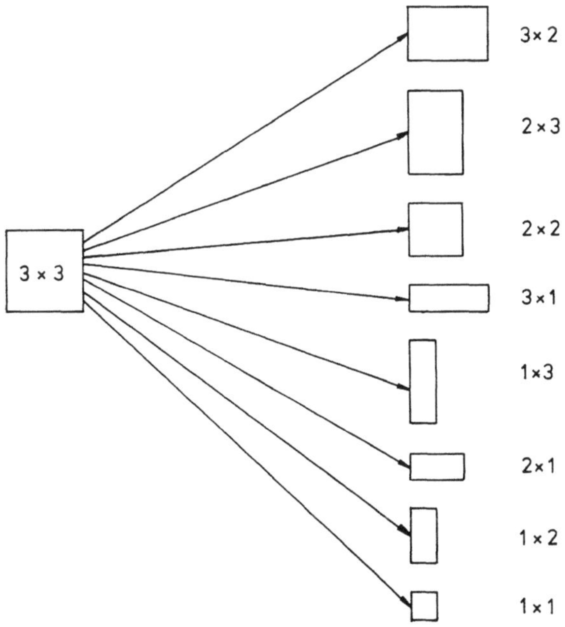

Fig. 10.3a. Partitions of the square 3×3: single-stage connection graph

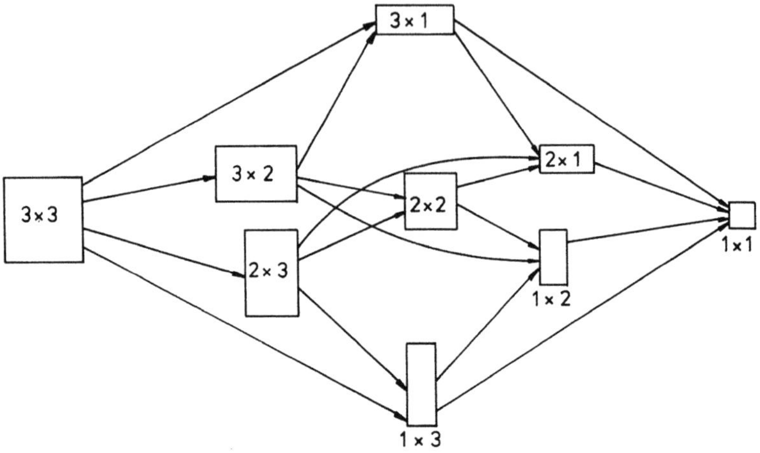

Fig. 10.3b. Partitions of the square 3×3 (with one-cuts): multi-stage connection graph

which the other eight rectangles $3 \times 2, \ldots, 1 \times 1$ result as potential output. (To illustrate this more clearly the "nodes" of the graph have been drawn as rectangles with corresponding measurements.)

If the single-stage process is deviated from and a multi-stage process used instead, in which the parts can be further cut down, whereby, however, a rectangle can only be divided into two smaller rectangles by means of a single guillotine cut, the result is the connection graph shown in Fig. 10.3b. This graph has a relatively complex structure in spite of the simple example.

If the size of a figure g is measured by its length, area or volume $V(g)$ – dependent on its dimensionality –, the following is directly true: $V(g) > V(h)$ if $g \langle Z \rangle h$. If g and h have the same convex form and the same orientation, i.e. they differ at the most in size, and if all imaginable divisions are possible in fact, the following is obviously valid:

(a) $V(g) > V(h)$ if and only if $g \langle Z \rangle h$;

(b) $V(g) = V(h)$ if and only if $g = h$.

These equivalences are especially true for all one-dimensional figures g and h. It is to be noted that two rectangles whose lateral lengths are in a different relationship do not have the same "form", and that substitution of the lateral lengths is only allowed when 90° turns are permitted.

The directed production connection between different types of goods gives a separation into four classes: Those goods-types which can be divided but which cannot be manufactured through the division of other types are known as primary factors. Those types which are manufactured through division, but which cannot be further divided, are the finished products. Those types which are manufactured through dividing and which in turn can be further divided are the intermediate products. Those goods which can neither be divided nor are manufactured through dividing are not involved and therefore play no part in the following discussion. (They form isolated nodes of the connection graph.) The foregoing considerations show that the primary factors are usually larger than the finished products. The largest figures are always primary factors (or not involved), the smallest are finished products (or not involved). Fig. 10.3 show clearly that the classification of a goods-type derives from the concrete production context and that the latter refers to the respective underlying decision-making situation.

Within a static model framework it is immaterial, as long as logical dependencies are borne in mind, whether the cutting processes of a planning period are actually to be run through simultaneously or consecutively because an abstraction is made in that all operations are projected on to a fictitious point in time and only the balance is taken into consideration. Intermediate products, which are always employed in the amounts in which they accrue and thus are not subject to any alterations in the periodical balance, can then be ignored. In such special cases, which in practice occur frequently, a division of the goods-types into two classes, primary factors on the one hand and finished products on the other, is possible. In general, however, it must be assumed (contrary to KOOPMANS 1951, p. 40) that intermediate products have to be taken into account.

A goods-type classification of another kind is based on the desirability of goods in the production planning framework as seen by the producer as decision-maker. Goods are regarded as desired when their output as products is desired, or when their employment as factors is to be avoided. In the opposite case they are described as undesired. Excess goods are those whose output or utilization in a production process is regarded neutrally (compared with the desired and undesired goods). Undesired and excess goods together form the group of the not (explicitly) desired goods. In relation to the output of a production process the excess products represent surplusage, undesired products are waste, and both, the non-desired products, are the remainder. In cutting processes the remainder or the waste, respectively, is called trim loss. Products of a type which in the given quantitative framework of the production process are mainly desired, undesired or excess are described as main products, waste products or co-products. As factors they are called by analogy main, waste, or co-factors. If goods of one and the same kind are seen chiefly as desired, undesired or excess as products and as factors then they are called main goods, waste goods ("bads") or co-goods. Refuse is, for example, a waste factor in incineration. For a long time industry treated waste water, waste gas and waste heat as co-products. (For a more detailed explanation of the definitions see DYCKHOFF 1987. FANDEL, in Chapter 9 of this book, denotes waste as "quantities for disposal" and calls "waste" the remainder.) The case usually examined in production theory is that without waste and co-goods (KOOPMANS 1951, pp. 59; FANDEL 1987). For the sake of simplicity a general classification of the goods-types into three classes is assumed here.

(G4) Desirability of goods-types independent of quantity

The assessment of the goods with regard to their desirability is the same for all goods of a particular type, i.e. with one and the same figure. The goods-types of the set G can thus be divided into three (disjunct and possibly empty) classes G_H, G_A, and G_N:

- the main goods (-types) G_H ,
- the waste goods (-types) G_A ,
- the co-goods (-types) G_N .

2. Characteristics of the Production Process

The real number z_g stands for the net output of the goods-type g in a production process, here a cutting process. $z_g < 0$ shows that on balance more of g is utilized than produced. If the goods-types in G are numbered by $1, \ldots, n$, a cutting process is represented by the activity

$$z = (z_1, z_2, \ldots, z_g, \ldots, z_n)' .$$

Not every vector of this type corresponds to a possible cutting process, i.e. belongs to the technology under discussion. In order to determine this we first restrict considerations to particular elementary cutting processes, the "partitions". For example,

$$z^1 = (0, 1, 1, 1, 0, 0, 0, 0, -1)'$$

describes the partition of an input object with length 9 into one part each of the lengths 2, 3, and 4, in the event that $g = 1, \ldots, 9$ symbolizes one-dimensional figures of lengths $V(g) = g$.

(a) Partitions

(Z0) The partitioning property

An activity (for the figures of G) is a partition when exactly one object with a figure in G is utilized and (at least) one geometrical cutting pattern for this figure exists in such a way that the resulting parts correspond in their figures exactly to the output objects of the activity (and can be imagined to be fictitiously reassembled spatially so as to correspond to the original input figure). The activity is a (technically) possible partition, i.e. it is an element of the underlying technology, when it is a partition with its cutting pattern satisfying given technical, organizational or alike restrictions.

Z0 has the character of a definition whereby the notions "cutting pattern", and "technical, organizational or alike restrictions" remain undefined, as an intuitive prior understanding is relied upon. A general, formal and precise definition hardly appears possible, at most in special cases, e.g. for one-dimensional partitions without further restrictions (see Proposition 1 below). There are, however, certain characteristics which can be formulated and which all activities must necessarily possess to be partitions. One of these characteristics is the unavoidable existence of more than one output object.

(Z1) Elementary diverging process

Each partition is an "elementary diverging process", that is to say, an activity z with a single input object and several output objects. This means that there is a goods-type (figure) g of G with $z_g = -1$ as well as one or more goods-types h of G with $z_h > 0$, whereas $z_f = 0$ for all other goods-types f of G.

Partitions transform a unit of quantity of goods-type g (input) into quantities z_h of other goods-types h (output); that is to say, they are diverging processes with z_h as yield coefficients. A partition is a rigid joint process provided that at least two yield coefficients are positive. The following characteristic renders the premise G3 in terms of formal requirements for partitions as activities.

(Z2) Non-cyclic production

If the goods-types g and h are input and output respectively, in a possible partition z (i.e. $z_g < 0$, $z_h > 0$), then $g \langle Z \rangle h$ must be valid for the non-cyclic relation $\langle Z \rangle$ of the production connection existing according to G3, i.e. h is geometrically contained in g as a possible result of a division.

While Z1 and Z2 hold true for a wide spectrum of non-cyclic joint processes the fact that the yield coefficients display the frequency of the parts in partitions is more limiting.

(Z3) Integer yield cofficients

If z is a partition then all the activity coefficients z_g for the goods-types g out of G are integers.

Z3 creates a clear difference to the usual approaches (though not necessarily to the results) of production theory in the analysis of elastic joint production (cf. STACKELBERG 1932; GERHARDT 1966; WITTMANN 1968; MÜLLER-MERBACH 1981).

(Z4) Volume conservation

If $z = (z_1, \ldots, z_n)$ is a partition (with due regard to all resulting parts) then

$$z_1 \cdot V(1) + \ldots + z_n \cdot V(n) = 0$$

for the goods-types $g = 1, \ldots, n$ and their volumes $V(g)$.

The equation in (Z4) is a special case of a quantity balance as is regularly to be found in input/output models (cf. KNOLMAYER 1980; MÜLLER-MERBACH 1981). Here, unavoidable trim losses due to the cutting technology occurring in practice are either ignored or have to be considered as "parts" with certain figures g of G. Obviously valid is

Proposition 1: If the goods-types g of G are one-dimensional figures (within the meaning of G1 – G3), a partition is uniquely defined by an activity $z = (z_1, \ldots, z_n)$ which satisfies the conditions Z1, Z3, and Z4.

This special feature of one-dimensional cutting processes compared with multi-dimensional processes is based upon the fact that every joining together of the parts of a one-dimensional input object results in principle – i.e. when no further limitations obtain – in a possible pattern for the realization of a partition. In general, however, conditions Z1 – Z4 are necessary, but not sufficient for a complete characterization. This can be seen when the "partition" of a cube with lateral length 2 into two spheres of volume 4 each is to be displayed by means of the activity $(-1, 2)$. This sort of activity is for geometrical reasons impossible in a theory of pure cutting processes (that is, without distortions!). A sufficient formal characterization of partitions is, however, possible in special cases, even where more dimensions occur, e.g. in the important practical case of two-dimensional multi-stage guillotine cutting patterns (DYCKHOFF 1987).

A very large number of technically possible partitions, reaching into the millions, is typical for cutting problems in practice (e.g. in the case of HEICKEN and KÖNIG 1980; cf. GILMORE and GOMORY 1963, p. 865, as well as Section III.2 of this chapter). The following postulates are thus practically not problematic and should exclude all unusual cases from the theoretical considerations.

(Z5) Conditions of existence

(a) There exists at least one possible partition with (at least) one main product as output.
(b) There is no partition in existence by means of which a waste or co-factor is transformed into main or co-products only.

(b) Cutting Processes

Cutting processes come about through a combination of partitions. The technically possible partitions form the "basis" of the technology, which is why they are called basic activities.

(A0) Existence of a finite number of basic activities

There exists a non-empty finite set of non-trivial basic activities $z^1, \ldots, z^t$, all distinct from one another, which are defined naturally by the given technological framework and out of which all the other activities of the technology can be combined.

In his general (linear) activity analysis KOOPMANS (1951, p. 36) takes the basic activities as given. For the special case of cutting processes considered here the conditions Z0 – Z5 are to be understood as an attempt to put A0 into concrete terms. "Non-trivial" are all activities except the zero activity $z = \mathbf{0} = (0, 0, 0, \ldots, 0)$, i.e. the origin of the space of all goods-types involved. This "activity" says that on balance no alterations to the goods take place, just as if nothing has been done. It is proper to regard this borderline case as a possible activity. It does not, however, count as a basic activity.

(A1) Inaction possible

The zero activity $\mathbf{0}$ is a (trivial) cutting process which is always possible.

Basic activities must not only be possible in principle but also within the given technical or organizational conditions. There are in principle 49 different partitions cutting a square with lateral length 3 into rectangles with integer lateral lengths and fixed orientation (cf. Fig. 10.3a). As can be seen in Fig. 10.4a one and the same partition can result from cutting patterns differing from each other in a non-trivial way. If for technical reasons only guillotine cuts are allowed the pattern shown in Fig. 10.4b is the only one of the 49 which cannot be realized. Fig. 10.4c illustrates two four-stage guillotine patterns.

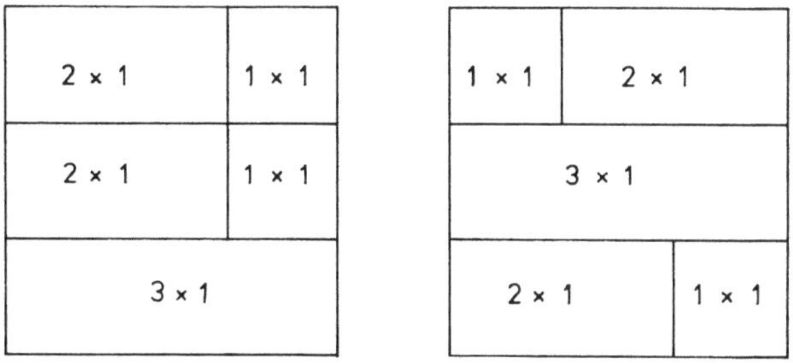

Fig. 10.4a. Different patterns of the same partition

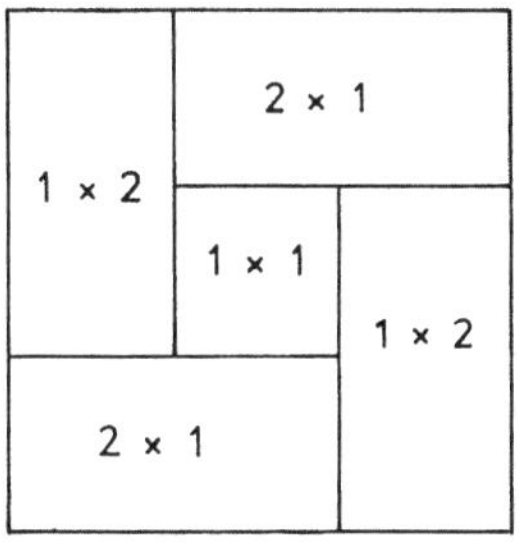

Fig. 10.4b. Nested pattern

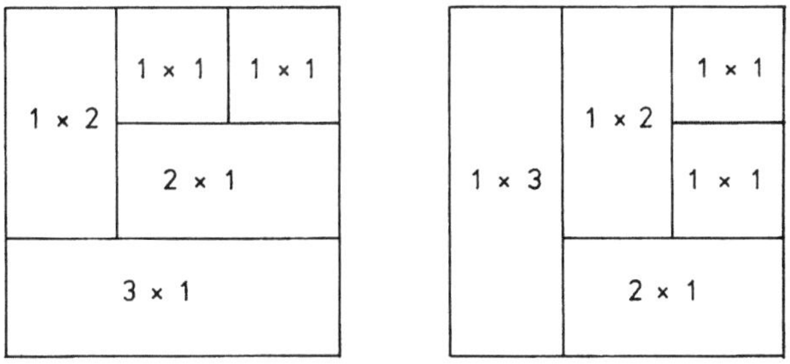

Fig. 10.4c. Four-stage guillotine patterns

If in extreme cases the guillotine cuts can only be made with a single blade, then there are only two possible partitions of a square 3×3 – on the one hand into the parts 3×1 and 3×2 and on the other hand into 1×3 and 2×3, both with the corresponding yield coefficient 1. If it is accepted that the parts thus arising can, as intermediate products, be further divided, there is a total of 12 basic activities. They are illustrated by the "input/output graph" (MÜLLER-MERBACH 1981) of Fig. 10.5 which is an extension of the corresponding connection graph (Fig. 10.3 b) by the yield coefficients of the partitions.

It is reasonable that the combination of partitions should mean their addition. For example, starting with 11 bars of length 9, 9 bars of length 5, and 15 of

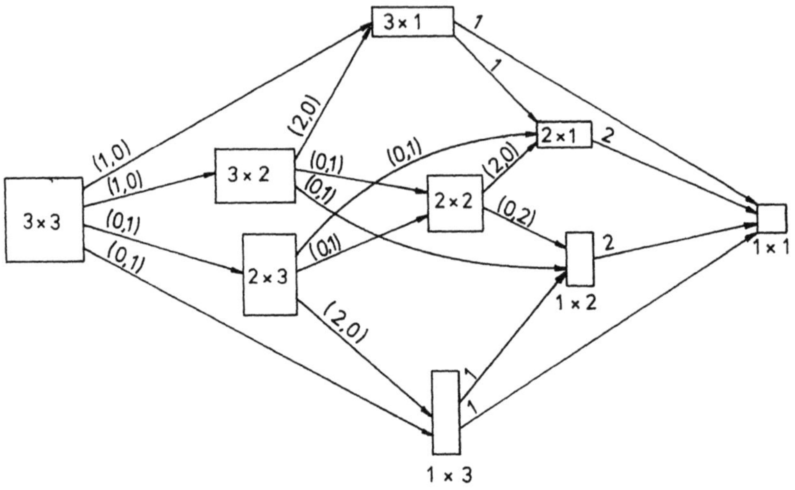

Fig. 10.5. Partitions of the square 3×3 with one-cuts: input/output graph

length 3 can be got by dividing each of 9 input bars into one part each of the lengths 5, 3, and 1, as well as 2 input bars into three parts each of length 3:

$$\begin{aligned} & 9 \cdot (1,0,\ 1,0,1,0,0,0,\ -1) \\ + & 2 \cdot (0,0,\ 3,0,0,0,0,0,\ -1) \\ = & \quad (9,0,15,0,9,0,0,0,-11)\ . \end{aligned}$$

The remainder consists of 9 bars of length 1. A cutting process without a remainder using only 10 input bars:

$$(0,0,15,0,9,0,0,0,-10)$$

is impossible as it cannot be combined from suitable partitions.

(A2) Additivity

If a and z are (technically) possible cutting processes, then their sum $a+z$, i.e. the sum of the quantities of each of the goods-types involved, is also a (technically) possible cutting process.

According to A2 an activity can be utilized many times ("added to itself") whereby the input and output amounts multiply. In combination with A1, A2 implies:

(A2a) Repetition possible

If z is a possible cutting process then so are all activities $a = v \cdot z$ for positive integer repetition factors $v = 1,2,3,\dots$

(A2b) (Additive) Combination possible

The combination

$$a = v_1 z^1 + \dots + v_t z^t$$

of possible cutting processes $z^1, \ldots, z^t$ by means of non-negative, integer repetition factors $v_1, \ldots, v_t$ results in a further possible cutting process.

Hence, by supposing A0 – A2 and allowing zero to be a repetition factor all possible cutting processes can be described as (additive) combinations of the basic activities. The matrix

$$T := (z^1, \ldots, z^t)$$

defined by the basic activities is therefore called the technology matrix. It is minimal when the technology can no longer be fully combined ("spanned") from the remaining basic activities after a single basic activity has been removed. In other words every possible activity z can be described as the product of the technology matrix T and a suitable vector $v = (v_1, \ldots, v_t)'$ of repetition factors:

$$z = T \cdot v \ .$$

As has been said already, geometrically defined goods are not "arbitrarily divisible" in the usual sense. Therefore, non-integer repetition factors (also called "activity levels") seem to make no sense. They are, however, appropriate in two frequently occurring cases: Firstly where there is a high number of pieces by which a continuous approximation of the actually discrete repetition factors usually appears to be justified. The second case concerns the so-called "1.5-dimensional" cutting problems (and by analogy the "2.5-dimensional" ones; cf. Chapter 12 of this book). They can be seen as originally two-dimensional problems which are degenerated to one-dimensional ones because for one of the two dimensions an actual cutting problem no longer exists by reason of "arbitrary divisibility". This applies to the cutting up of rolled goods when several rolls of the same width (and quality etc.) do not have to be differentiated from a single roll of the same width whose length is the same as the total length of all the single rolls. The lengths of the individual rolls can mostly take on arbitrary values. A basic activity then specifies the partition of a roll of a given standard length into part rolls of narrower widths. Rolls of an arbitrary length can also be cut up according to the same pattern into rolls of shorter lengths or of a smaller diameter by uncoiling (and cross-cutting) so that in such cases the assumption of "arbitrary divisibility of the activity levels" (also called "proportionality" or "linear homogeneity") is justified. Although some of the following results are valid without this assumption (DYCKHOFF 1987) it is generally supposed here for the sake of simplicity.

(A3) Arbitrarily divisible activity levels

If z is a possible cutting process then so are $a = v \cdot z$ for all positive numbers v as repetition factors (activity levels).

3. Characteristics of the Technology

A0 – A3 define linear technologies as introduced by KOOPMANS (1951). Furthermore, he postulated three regularity conditions. Here they are already implied through conditions Z1 – Z5, partly in a more general version taking waste and co-goods into account.

(A4) Irreversibility of production

The zero activity cannot be described as a proper combination of activities; hence, from $a+z=\mathbf{0}$ follows $a=z=\mathbf{0}$.

Irreversibility excludes among others a technology which consists of the whole goods-space. It is a consequence of the assumptions G3 and Z2 concerning the non-cyclic nature of the production. The goods, defined as geometric objects, can be divided but not joined together. Production processes run in definite directions and cannot be reversed (at least not without additional expenditure).

(A5) Possibility of production

There exists at least one possible activity z with positive output of a main product ($z_g > 0$ for a g of G_H) or with positive input of a waste factor ($z_g < 0$ for a g of G_A).

If only desired goods are explicitly taken into account, i.e. they are all main goods, A5 reduces to the usual postulate that output is possible, at all. This is, however, already implied through A0 (or Z5) and Z1, according to which at least one basic activity exists producing as an elementary diverging process (more than) one output object. According to A5 output does even not necessarily have to be possible if in its stead waste input is possible (e.g. refuse incineration). This case cannot occur here because Z5a guarantees the existence of a technically possible partition with one type of main product as output.

KOOPMANS' (1951) postulate: No output without input, rules out the possibility of a "Land of Cockaigne" for the case of only desired goods. A Land of Cockaigne, however, would exist if main products were produced solely from waste.

(A6) Impossibility of the Land of Cockaigne

If z is a possible activity without input of main goods and without output of waste then z is also without output of main goods and without input of waste (i.e. there is no possible activity with $z_g \geq 0$ for all g of G_H and $z_g \leq 0$ for all g of G_A as well as $z_g > 0$ or $z_g < 0$ for at least one g of G_H or G_A, respectively).

A6 follows from Z1 and Z2 in combination with A0 – A3 and Z5b. For, there are input goods for every basic activity, because of Z1. Even in multi-stage processes because of their non-cyclic nature (Z2 and G3), there is always at least one goods-type which remains as primary factor. Z5b prevents waste or co-goods from being primary factors if only main and co-products are manufactured.

To sum up the following can be asserted: The properties Z1 and Z2 together with A0 – A3, which imply non-cyclic elastic joint production, already contain postulates A4 – A6 if no waste or co-products have to be taken into account; otherwise Z5 has to be presupposed in addition.

III. Efficiency Considerations

Technologies consist of activities and describe the fundamental production possibilities. Not all technically possible activities are, however, economically relevant. A minimum requirement is the economic principle which an activity must follow: The quantity of a factor of production can only be decreased, or the quantity of a product only be increased, if quantities of other factors were increased and/or quantities of other products were decreased. This is, however, only valid in the usually discussed case of production without waste and co-products.

1. The Economic Principle in Case of Surplus and Waste

In activity analysis, the economic principle is based on the concept of efficiency of activities. The efficiency border of a technology defines the corresponding production function (cf. FANDEL 1987 and Proposition 3 below). By resorting to the formal concept of efficiency in the theory of multiple objectives, waste and co-goods can also be included. For this purpose the net output of a main good is regarded as an objective to be maximized and the net output of a waste good (a "bad") as an objective to be minimized, while the amounts of co-goods involved are disregarded (e.g. oxygen of the air in incineration processes). Then, a possible activity z is called efficient (with respect to the underlying technology) if every possible activity a dominating z is identical with z, with the eventual exception of the co-goods g of G_N, i.e. if there is no activity in the technology which strictly dominates z. Here, z is (weakly) dominated by a when

$$a_g \geq z_g \quad \text{for all main goods-types } g \text{ of } G_H \text{, and}$$
$$a_g \leq z_g \quad \text{for all waste goods-types } g \text{ of } G_A .$$

Strict dominance is found when in addition

$$a_g \neq z_g \quad \text{for at least one } g \text{ of } G_H \text{ or } G_A .$$

Without waste and co-goods the definition reduces to the usual efficiency concept in activitiy analysis and production theory. It can be interpreted as if all the goods not under consideration are co-goods. This means that the corresponding coefficients of the activities are not made explicit, so that only those reduced activities are observed. More generally, through projection into the less-dimensional space of the main and waste goods, the activities z in the goods-space (and with them the corresponding technology) can be reduced to activities $\tilde{z} = P(z)$. If in addition the sign of the waste quantities is changed by the projection a simplified formulation is obtained which, moreover, leads the above definition of efficient activities in the space of all goods-types back to a usual formulation for the sub-space of main and waste goods. Thus, if z is an activity of the technology as regards the total goods-space, the reduced activity $\tilde{z} = P(z)$ in the sub-space of main and waste goods is defined by

$$\tilde{z}_g := z_g \quad \text{for all } g \text{ of } G_H ,$$
$$\tilde{z}_g := -z_g \quad \text{for all } g \text{ of } G_A .$$

For the activities a and z of a technology and the corresponding reductions $\tilde{a} = P(a)$ and $\tilde{z} = P(z)$ follows: a dominates z if and only if $\tilde{a} \geq \tilde{z}$ (i.e. $\tilde{a}_g \geq \tilde{z}_g$ for all g of G_H and G_A); a dominates strictly for $\tilde{a} > \tilde{z}$ (i.e. $\tilde{a} \geq \tilde{z}$ and $\tilde{a} \neq \tilde{z}$).

Although the co-goods are neglected for efficiency considerations the question as to whether a reduced activity is actually technically possible can in general only be answered through stating a normal activity of the technology of which the reduced activity is a reduction. Therefore it is hardly possible to totally neglect the co-goods when characterizing partitions by Z1 – Z4. They usually correspond to the residual pieces resulting from the construction of the cutting patterns.

While the general assumptions A0 – A4 can be transferred directly to the reduced activities simplified formulations arise for Z5b, A5, and A6. Z5b states that there is no reduced basic activity with only non-negative coefficients. According to A5 it is impossible that all reduced activities of the technology consist of non-positive coefficients only. Because of A6 no reduced activity with $\tilde{z} > \tilde{0}$ exists. The reduced activities display the usual properties of the activity analysis without surplus and waste. Hence, A0 – A6 induce in the reduced space of main and waste goods the analogous properties which KOOPMANS (1951) postulates for main goods only. Consequently, some characteristics of technologies of pure cutting processes can be deduced from the characteristics of more general linear technologies, without having to have recourse to cutting-specific assumptions.

Proposition 2 ("Price Theorem"; cf. KOOPMANS 1951, pp. 64): Let $p = (p_1, p_2, \ldots, p_n)'$ denote a "price system" and the product $p' \cdot z$ be the "profit". Because of assumptions A0 – A3 a technically possible activity z is efficient if and only if a price system p exists with positive prices for the main goods, negative prices for the waste goods and zero prices for the co-goods so that the profit $p' \cdot z$ is maximum with respect to all activities of the technology. Then, $p \cdot z = 0$ as well as $p' \cdot z^j = 0$ for all basic activities z^j which can be combined to z with a positive repetition factor v_j.

The definition of a proper combination of activities for the case when all repetition factors of these activities are positive yields the following

Corollary:

(a) If a proper combination of basic activities is efficient, then so are all these basic activities.
(b) If any proper combination of efficient basic activities is efficient, then so are all combinations of these basic activities, even improper ones.

According to A6 and in view of the zero activity (A1) an efficient activity always exists. A5 and A6 guarantee furthermore that at least one basic activity is efficient (DYCKHOFF 1987).

Proposition 3 (cf. HILDENBRAND 1966): Because of assumptions A0 – A3 there exists a continuous production function f_T. That is to say, f_T is a real-valued function defined on the goods-space with $f_T(z) = 0$ if and only if z is efficient (and technically possible).

Only in special cases, the (implicit) production function of a technology can be described in a closed analytic form. At the most this can be done piecewise for different parts of the goods-space. Instead of this, in the following it will be tried step by step in a constructive manner to restrict considerations from the whole technology to smaller subsets which contain the efficiency border, and where possible, hardly any other activities.

2. Efficiency of Partitions

Because all efficient activities, except the zero activity, can be described as a proper combination of efficient basic activities (see Corollary (a) of Proposition 2) it is sufficient to characterize the efficient partitions on the one hand and on the other the efficient combinations of efficient partitions. As far as the latter is concerned statements for cutting processes which clearly go further than those of the general activity analysis hardly seem possible. The following investigation therefore concentrates on the characterization of efficient partitions.

A technically possible basic activity is called effective when it is not dominated by another possible basic activity or by a multiple of a possible basic activity. Trivially, every efficient basic activity is effective. An elementary diverging process, given as a basic activity according to Z1, with just one input good is effective if and only if no other basic activity with the same input exists by which the output of at least one main product is higher, or of one waste product is lower, without the output of at least one main product sinking, or of a waste product rising, at the same time. (Besides, every effective basic activity, defined according to Z1, is also efficient, if A3 is not valid, i.e. if the activity level is not arbitrarily divisible.)

Therefore, effective partitions can be strictly dominated only by a proper combination of at least two other partitions. Conversely, every efficient activity, apart from the zero activity, is a proper combination of effective basic activities. Hence, it is sufficient to limit considerations to the effective partitions and combinations thereof. It leads to an improvement when there are many non-effective basic activities which can thus be excluded from consideration. This usually holds true. For example, of the 29 possible partitions of length 9 into integer part-lengths, only 7 are effective in the case that the lengths 9, 4, 3, and 2 are main goods and all other lengths co-goods. They are characterized by having at most one remaining piece of length 1. If instead of 4, 3, and 2 the lengths 6, 5, and 4 are the main products, only the 5 partitions shown in Table 10.1 are effective.

The first three partitions are only differentiated as regards the amount of co-products accruing. In the space of the main goods of the lengths 9, 6, 5, and 4 all three have the same reduced form: $\tilde{z} = (-1, 1, 0, 0)'$. In practice it is usually of no consequence how the co-products are composed; often only their volume is of interest. In the three partitions this "trim loss volume" always amounts to three units of length. Partitions which display the same reduced form in the subspace of main and waste goods, hence only differing with respect to the co-goods, are called equivalent. Partitions equivalent to effective partitions are

Table 10.1. Effective partitions of the length 9 into the main product lengths 6, 5, and 4

Product length	Product frequency in effective partition No.				
	1	2	3	4	5
Main products					
6	1	1	1	0	0
5	0	0	0	1	0
4	0	0	0	1	2
Co-products					
3	1	0	0	0	0
2	0	1	0	0	0
1	0	1	3	0	1

Table 10.2. Strictly effective partitions of the input length 100 into the main product lenghts 35, 19, and 13

Main product lengths	Product frequency in strictly effective partition No.											
	1	2	3	4	5	6	7	8	9	10	11	12
35	0	0	0	0	0	0	1	1	1	1	2	2
19	0	1	2	3	4	5	0	1	2	3	0	1
13	7	6	4	3	1	0	5	3	2	0	2	0
"Trim loss"	9	3	10	4	11	5	0	7	1	8	4	11

themselves effective. They have the same surplus volume (and of waste and main products, respectively). It must, however, be stressed that equivalent partitions can certainly be quite different from another as regards technical, organizational or other restrictions.

Equivalent partitions of a technology can thus be identified with respect to efficiency. Let one of them be chosen as a representative and denoted as strictly effective, namely, the one which displays the lowest number of co-products. In Table 10.1 this is partition No. 1. In the case of one-dimensionality without further restrictions, strictly effective partitions always result in only one part of a co-product as remainder length. If in addition there are no waste products the remainder-length must be shorter than the smallest main product length. This does not necessarily apply if waste exists or where technological restrictions are given or, analogously, for more dimensional partitions. Therefore, the following statements concentrate on one-dimensional partitions without further restrictions and without waste. (For the inclusion of waste under weak additional assumptions, see DYCKHOFF 1987.) It is then sufficient to limit the consideration to strictly effective partitions.

If only main goods are to be found all possible partitions are strictly effective. In the one-dimensional case without further restrictions and without waste the strictly effective partitions can be determined by means of a simple algorithm using (limited) lexicographic enumeration. For example, Table 10.2 displays the

12 strictly effective partitions, in lexicografically ascending order, for the input length 100 and the main product lengths 35, 19, and 13 (with otherwise co-products, only).

On the other hand there are about 200 million different possible partitions for the input length 100. Denoting the number of all possible partitions of the length s by $N(s)$, we have: $N(4) = 4$, $N(9) = 29$. The greater the lengths the quicker the number $N(s)$ increases (ANDREWS 1976, p. 2), e.g.

$$
\begin{aligned}
N(10) &= 41 = 4.1 \cdot 10^{1} \ , \\
N(20) &= 626 \approx 6.3 \cdot 10^{2} \ , \\
N(50) &= 204225 \approx 2.0 \cdot 10^{5} \ , \\
N(100) &= 190569291 \approx 1.9 \cdot 10^{8} \ , \\
N(200) &= 3972999029387 \approx 4.0 \cdot 10^{12} \ , \\
N(1{,}000) &\approx 2.4 \cdot 10^{31} \ , \\
N(10{,}000) &> 10^{106} \ .
\end{aligned}
$$

The logarithmic form of an approximation formula for $N(s)$ for large s, which can deliver quite good values already from $s = 20$ onwards, is as follows:

$$\log_{10}(N(s)) \approx 1.114 \sqrt{s} - \log_{10}(s) - 0.84062 \ .$$

Insofar as, with one-dimensional cutting problems, the lengths involved are changed through the choice of a suitable gauge into the smallest possible integers, the lengths of the objects which have to be divided are in practice seldom less than 100 and can be as much as 1000 or even higher. The number of possible partitions is thus usually very large. The number of effective and strictly effective partitions is indeed only a fraction of all possible partitions but can also be very large and often is. There are in general no formulae which, analogously to what has been stated above, permit the direct, simple determination, even approximately, of the number of effective or strictly effective partitions. For a concrete cutting problem in the paper industry (GILMORE and GOMORY 1963) where an input "length" $s = 872$ (= paper rolls 218 inches wide with 1/4 inch exactness) has to be divided into 30 different "order-lengths" (narrower rolls as the main products) in the area from 86 up to 324, "only" about 100,000 (= 10^5) strictly effective partitions out of possible 10^{29} exist. GILMORE and GOMORY (1963, p. 865) point out that already with 40 order-lengths (as main products) of similar measurements "the number of cutting patterns [strictly effective partitions; H.D.] can easily exceed 10 or even 100 million. Problems of these sizes are often encountered in practice." In the case study of HEICKEN and KÖNIG (1980) where tubes of lengths $s = 98$ (bore-holes) are to be cut up into order-lengths of between 46 and 5 (bore-holes) there are still more than one million strictly effective partitions out of more than 150 million which are possible, in principle.

A further reduction of the set of partitions being relevant with respect to efficiency is possible if strictly effective partitions are dominated by combinations of other partitions. One can proceed analogously to the combination of Leontieff processes (cf. FANDEL 1987) only that partitions are inversely related to the latter. A partition in cutting up material objects (elementary diverging process) is namely the opposite to the bill of material in the assembly of material objects (elementary converging process) whereby the cutting pattern corresponds to the

Table 10.3. Efficient partitions of the input length 100 into the main product lengths 35, 19, and 13

Main product lengths	Product frequency in partition No.						
	1	2	6	7	9	11	12
35	0	0	0	1	1	2	2
19	0	1	5	0	2	0	1
13	7	6	0	5	2	2	0
Trim loss volume	9	3	5	0	1	4	11

production drawing (cf. the typology of input/output processes in MÜLLER-MERBACH 1981). In the example shown in Table 10.2 partition No. 5 is strictly dominated by a combination of partitions No. 1 and 6 and thus is not efficient:

$$\begin{aligned} & 0.2z^1 + 0.8z^6 \\ & = 0.2 \cdot (-1,0,0,7) + 0.8 \cdot (-1,0,5,0) \\ & = (-1,0,4,1.4) \; . \\ & > (-1,0,4,1) = z^5 \end{aligned}$$

(The tilde for denoting the reduced activities has been left out.) Correspondingly it can be seen directly in Table 10.2 that partition No. 4 is identical with a balanced combination of the two partitions No. 2 and 6 and can be struck out because of (weak) domination, although it may possibly be efficient. (This may in certain circumstances be disadvantageous as regards the costs of refitting the cutting machinery on changing patterns, something which is neglected in this production-theoretical approach!) In the end, of the 12 partitions of Table 10.2 only 7 efficient ones remain, according to Table 10.3.

IV. Production Models

In general it is hardly possible, or at least too costly, to limit the technology matrix T to partitions in such a way that the general production model

$$(1) \qquad z = T \cdot v \; , \qquad v \geq \mathbf{0} \; ,$$

describes exclusively efficient parts of the technology. On the other hand the multitude of possible partitions makes it impracticable to display the whole set of possible productions. A way out of the dilemma is to formulate the matrix T so that the production model on the one hand is accessible for practical purposes to the production management and on the other includes all the relevant parts of the technology. In certain cases the method of "delayed pattern generation" (GILMORE and GOMORY 1961, 1963, 1965) allows the explicit formulation of the matrix T to be dispensed with. In this method, the partitions are implicitly taken into account using a vertical decomposition approach (as opposed to the "horizontal decomposition" by DANTZIG and WOLFE 1960) in the search for the optimal cutting process.

In the following there will be presented two model approaches specifying the general form (1) for cutting processes. The cutting models in the literature and also the method of delayed pattern generation are based almost exclusively upon the first approach. Two sets of data, which are known in the practice of production planning for cutting processes, form the starting point for the construction of the matrix T, from now on called production matrix:

D = order figures: those goods-types desired as output (main products); mostly those figures for which demand exists;
S = stock figures: goods-types available as primary factors from which parts of the desired output can be produced.

Without restricting generality it can be supposed that desired figures can in principle be produced by the partition of at least one of the available figures and that no demand for primary factors exists (any longer or has already been met) so that the intersection of both sets is empty ($D \cap S = \phi$). The set D is not necessarily identical with the set of all desired goods. In particular stock figures can belong as factors to the main goods. Furthermore the following sets of goods-types (figures) are of importance:

G = relevant figures: those goods-types basically involved in the cutting process, i.e. figures which as input or output (and perhaps as intermediate products as well) have to be taken into account. Goods-types neither being input nor output are of no importance for the cutting process and do not need to be considered explicitly; i.e. rows of matrix T having only zeros as elements can be deleted.
I = input figures: those goods-types which might be input, i.e. figures which can be divided in the course of the production process. They correspond to rows of matrix T which display -1 as an element at least once.
K = output figures: those goods-types which might be output, i.e. figures which in the course of the production process can be manufactured as parts (in partitions). They correspond to rows of matrix T with positive elements.
R = trim loss or remainder figures: those goods-types which might be output without being (directly) desired, i.e. figures which do not correspond to an order figure and in which, in the light of the considered partitions, residual pieces might possibly accrue.

Due to definition, $G = I \cup K$ and $K = D \cup R$ with $D \cap R = \phi$. Even with the same order figures D and stock figures S production models may differ from one another with respect to the four sets G, I, K, and R, namely because of different partitions as underlying basic activities. Furthermore, it is assumed that waste products, if they occur at all, cannot be further divided into main or co-products, and also do not exist as primary factors (stock figures).

Both fundamental model approaches can be interpreted as analogies to two (linear programming) solution approaches for the choice between different processes in multi-stage, multi-product manufacturing, in German well known as "Alternativkalkulation" (alternative calculation) and "arbeitsgangweise Kalkulation" (stepwise calculation).

The "alternative calculation" regards the different process combinations through all the production stages as imaginary units and defines them as basic activities. Fictitiously, it is supposed a single-stage production process with the result that pure intermediate products do not appear explicitly. Therefore, only the stock figures come into consideration as input figures, i.e. $I = S$. For every input goods-type of S, partitions are determined (strictly) effective with respect to the order figures D, whereby possibly existing waste goods have to be taken into account in a suitable manner. Dominated partitions may be eliminated. The partitions established in this way define as basic activities the production matrix T in (1) and thus the (single-stage) production model of the alternative calculation.

A more specific model formulation is retained via double-indexing the partitions defining T, with an index i for the input (or stock) figures $I(=S)$ and a running index j for the partitions of i. Further, let be

- J_i the set of relevant partitions $[i,j]$ for the figure i;
- v_{ij} the repetition factor of partition $[i,j]$, i.e. the frequency with which objects of figure i are divided according to a pattern conforming to $[i,j]$;
- b_{kij} the yield coefficient of partition $[i,j]$ as regards the goods-type k, i.e. the number of parts of figure k which accrue on the partition of an object of figure i according to the pattern $[i,j]$;
- x_g the (gross) input of goods-type g, i.e. the total number of divided objects of figure g;
- y_g the (gross) output of goods-type g, i.e. the total number of parts of figure g which accrue through partitions;
- z_g the net output of goods-type g, i.e. the number of objects of figure g which on balance result more from partitions than are divided themselves.

Along with the stock and order figures the unintentionally accruing joint products are also involved as goods-types in the cutting process, i.e. $G = S \cup D \cup R$. S and D are given data. The remainder figures R come about as a result of the partitions taken into account.

Basic model of alternative calculation:

(a) $$z_g = y_g - x_g \,, \qquad g \in G \,,$$

(b) $$x_i = \sum_{j \in J_i} v_{ij} \,, \qquad i \in S \quad (x_i = 0 \text{ otherwise}) \,,$$

(c) $$y_k = \sum_{i \in S} \sum_{j \in J_i} b_{kij} v_{ij} \,, \qquad k \in K \quad (y_k = 0 \text{ otherwise}) \,,$$

(d) $v_{ij} \geq 0$ (and integer if necessary), $i \in S$, $j \in J_i$.

The linear programming models found in the literature on trim loss or cutting stock optimization are almost without exception formulated in this way as single-stage models, even if the actual cutting process is carried out in several successive steps. In contrast, the approach of "stepwise calculation" is multi-stage in that the processing possibilities are explicitly formulated at every stage and the individual stages are combined by way of quantity balances. For cutting processes extremely simply structured partitions are assumed for each individual cutting stage where the stages are supposed fictitiously if the actual cutting process is single stage. These simple partitions are known as "one-cuts" (DYCKHOFF 1981; RAO 1976, in a paper unnoticed until now, called them "binary cuts"). In practice they appear naturally when the cutting machine has only one blade (cf. Chapter 11 of this book). It is crucial that now the input and output figures do not build disjunctive sets ($I \cap K \neq \emptyset$) because the output figures of preceding partitions represent the input figures of succeeding partitions. In contrast to the familiar model of stepwise calculation in converging manufacture there are no given exogeneous stage indices here. Their function is taken over by the indices of the intermediate figures (intermediate products); these are those figures (goods-types) which can be input as well as output and therefore define the set $L := I \cap K$. Because every intermediate figure can only arise through the partition of a larger object and can only be used for the production of smaller objects a certain order for the considered partitions and the intermediate figures is the result. In general this order is only partial. It is complete for one-dimensional cutting processes in that it results in a definite sequence of the intermediate figures which quasi correspond to stage indices. Nevertheless, the individual parts of the stock figures usually run through only a few of the "stages".

The stock figures now form a subset of the input figures ($S \subset I$). Only one-cuts are allowed as partitions. Through a series-connexion of one-cuts it is possible to reproduce each ever so complex cutting process. For, each partition can be imagined as being the result of the individual pieces being cut one after another firstly from the original input object and then from the resulting residual parts (even when this in reality should prove to be impossible). Not all one-cuts have to be included in the model. With one-dimensionality, strictly effective, non-dominated one-cuts are sufficient, i.e. those which display at least one order-length plus remainder-lengths as output. One-cuts have to be defined not only for the stock figures but also for all those encountered remainder figures which are relevant as (potential) input figures.

The symbols of the model of alternative calculation can be retained for a specific formulation of the model of stepwise calculation but have to be partially newly interpreted. This is essential for the sets I and J_i (with $i \in I$):

J_i – relevant one-cuts for the figure i (by which as a rule at least one order figure is produced),

I – input figures by reason of the given stock figures of S and those output figures which accrue successively as a result of one-cuts and which (as a subset of K) are suitable for further one-cuts.

Unless the (potential) input figures I and also the output figures K can be specified from the start by simple observations they have to be determined recursively: Firstly the relevant one-cuts J_i for all stock figures i of S are constructed. Relevant are especially those one-cuts by means of which order figures can be gained from the accruing parts through following one-cuts. By dominance considerations, several of the one-cuts can possibly be eliminated, either immediately or later. For one-dimensionality, things can be limited to those one-cuts by which exactly two pieces are cut, of which at least one has an order figure. Output figures found among the one-cuts so determined are then registered (to determine K) and those among them which can be used to provide still further order figures are added to the set of input figures. Further relevant one-cuts J_i are constructed for the new input figures, and so on. The output and input figures registered up to the end form, with D and S, the sets K and I.

Basic model of stepwise calculation:

(a) $z_g = y_g - x_g \,, \qquad g \in G \,,$

(b) $x_i = \sum_{j \in J_i} v_{ij} \,, \qquad i \in I \qquad (x_i = 0 \text{ otherwise}) \,,$

(c) $y_k = \sum_{i \in I} \sum_{j \in J_i} b_{kij} v_{ij} \,, \qquad k \in K \qquad (y_k = 0 \text{ otherwise}) \,,$

(d) $v_{ij} \geq 0$ (and integer if necessary), $i \in I$, $j \in J_i$.

Apart from the replacement of the set S of stock figures by the set I of input figures and the other meaning of the symbol J_i, the production model can be interpreted without fundamental change. On close observation, however, serious differences can be seen in the structures of both model approaches. Firstly, the input/output balances (a) now can in no circumstances be dispensed with because there are always intermediate figures which constitute the non-empty intersection L of I and K. Secondly, the input balances (b) and the partitions have to be formulated not only simply for the stock figures of S but for all potential input figures I. Thirdly, the partitions in question, i.e. the relevant one-cuts, have a much simpler structure in comparison with the partitions to be found in the (single-stage) alternative calculation approach. The tendency for practical problem sizes can be so recorded that when all efficient cutting processes can be described by both models, the (multi-stage) model of stepwise calculation shows clearly more balance equations but far fewer partitions with their repetition factors. In exceptional cases this does not have to apply, as the first of the three examples in Table 10.4 illustrates.

Table 10.4. Comparison of matrix sizes for three concrete examples

Example (s = largest stock length; m = number of order lengths)				
	Size of the production matrix			
Model	No. of rows	No. of columns	Nonzero elements	Matrix density (%)
1 ($s = 732$, $m = 12$, $s/m = 61$)				
single-stage	16	378	1,526	25.23
multi-stage	306	1,714	5,048	0.96
2 ($s = 872$, $m = 30$, $s/m = 29.06$)				
single-stage	31	52,721	273,411	16.73
multi-stage	529	12,300	33,565	0.51
3 ($s = 98$, $m = 42$, $s/m = 2.33$)				
single-stage	43	1,397,747	9,250,699	15.39
multi-stage	90	1,994	5,772	3.22

Table 10.4 gives some characteristic values of the production matrix T for three concrete examples of one-dimensional cutting problems adapted from the case studies in Chapter 11 (of this book), GILMORE and GOMORY (1963), as well as HEICKEN and KÖNIG (1980), without taking further problem specific restrictions into account. Rows for those remainder lengths which are shorter than the smallest order-length have not been included. The size of the "multi-stage" model (of stepwise calculation) can be further reduced with the help of some deeper considerations (STADTLER 1988; DYCKHOFF 1987). However, in terms of computing time and memory this approach seems to be inferior compared with the "single-stage" model (of alternative calculation) if implicitly used within the very efficient Gilmore/Gomory-method of "delayed pattern generation" (STADTLER 1988; contrary to RAO 1976), provided that this algorithm can be implemented without too much effort. This disadvantage does not affect the fundamental production-theoretical importance of the multi-stage model. In particular to be emphasized is the fact that the algorithm of KNOLMAYER (1980) for the elimination of balance equations in linear programming models allows the model of stepwise calculation as one extreme formulation to be reformulated step-by-step into the alternative calculation model as the other extreme. Thereby the conditions for the equivalence of both models and for all intervening model formulations are pointed out clearly (DYCKHOFF 1987).

Multi-dimensional cutting processes can also be described with the stepwise calculation model, although the size of the production matrix hardly makes it of interest for practical use. This can be shown through the simple example of cutting up squares with the measurements 3×3 into smaller rectangles (cf. Fig. 10.3). Order- and stock-lengths, respectively, let be

$$D=\{2\times2,\ 3\times1,\ 2\times1\}\ ,$$
$$S=\{3\times3\}\ .$$

All other figures are co-products and therefore trim loss. Strictly effective are only the following six partitions, out of a total of 94 possible partitions:

No.	Partition
1	$[3\times3:2\times2+3\times1+1\times2]$
2	$[3\times3:2\times2+1\times3+2\times1]$
3	$[3\times3:3\times1+3\times1+3\times1]$
4	$[3\times3:3\times1+3\times1+2\times1+1\times1]$
5	$[3\times3:3\times1+2\times1+2\times1+1\times2]$
6	$[3\times3:1\times3+2\times1+2\times1+2\times1]$

Hence, with $I=S$, $K=D\cup R$, $R=\{1\times3,\ 1\times2,\ 1\times1\}$ for the single-stage model the following applies:

Alternative calculation

(a)

$$z_{3\times3}=-x_{3\times3}\ ,$$
$$z_{2\times2}=y_{2\times2},\ z_{3\times1}=y_{3\times1},\ z_{2\times1}=y_{2\times1}\ ,$$
$$z_{1\times3}=y_{1\times3},\ z_{1\times2}=y_{1\times2},\ z_{1\times1}=y_{1\times1}\ ,$$

(b) $x_{3\times3} = v_1 + v_2 + v_3 + v_4 + v_5 + v_6$,

(c). $y_{2\times2} = v_1 + v_2$,
$y_{3\times1} = v_1 + 3v_3 + 2v_4 + v_5$,
$y_{2\times1} = v_2 \quad v_4 + 2v_5 + 3v_6$,
$y_{1\times3} = v_2 + v_6$,
$y_{1\times2} = v_1 + v_5$,
$y_{1\times1} = v_4$,

(d) $v_j \geq 0$ (and integer if necessary), $j = 1,\ldots,6$.

For the construction of the model of stepwise calculation it is first necessary to ascertain the relevant one-cuts along with the accompanying input and output figures. The connection graph and the input/output graph are here of assistance. Figs. 10.3a and 10.5, however, assume that only guillotine cuts are allowed. This is certainly not absolutely necessary for one-cuts, but otherwise there are possible intermediate figures which are not rectangles (cf. DYCKHOFF 1987). If things are limited to guillotine cuts the following one-cuts as well as input and output figures result (cf. Fig. 10.5):

$$I = \{3\times3,\ 3\times2,\ 2\times3,\ 2\times2,\ 3\times1\} ,$$
$$K = \{3\times2,\ 2\times3,\ 2\times2,\ 3\times1,\ 1\times3,\ 2\times1,\ 1\times2,\ 1\times1\} .$$

No.	One-cut	No.	One-cut
1	[3×3:3×2+3×1]	5	[2×3:2×2+2×1]
2	[3×3:2×3+1×3]	6	[2×2:2×1+2×1]
3	[3×2:3×1+3×1]	7	[3×1:2×1+1×1]
4	[3×2:2×2+1×2]		

As can easily be seen, all the above six strictly effective partitions can be combined from this seven one-cuts, e.g. partition No. 2 from the one-cuts No. 2 and 5. Apart from No. 2, all the one-cuts generate at least one part with desired measurements (order figure). Without one-cut No. 2, however, (and subsequently No. 5) the partitions No. 2 and 6, in which the rectangle 1×3 appears, cannot be described.

Stepwise calculation

(a) $z_{3\times3} = -x_{3\times3},\ z_{3\times2} = y_{3\times2} - x_{3\times2},\ z_{2\times3} = y_{2\times3} - x_{2\times3}$,
$z_{2\times2} = y_{2\times2} - x_{2\times2},\ z_{3\times1} = y_{3\times1} - x_{3\times1}$,
$z_{1\times3} = y_{1\times3},\ z_{2\times1} = y_{2\times1},\ z_{1\times2} = y_{1\times2},\ z_{1\times1} = y_{1\times1}$,

(b) $x_{3\times3} = v_1 + v_2$,
$x_{3\times2} = v_3 + v_4$,
$x_{2\times3} = v_5$,
$x_{2\times2} = v_6$,
$x_{3\times1} = v_7$,

(c) $y_{3\times2} = v_1$,
$y_{2\times3} = v_2$,
$y_{2\times2} = v_4 + v_5$,
$y_{3\times1} = v_1 + 2v_3$,
$y_{1\times3} = v_2$,
$y_{2\times1} = v_5 + 2v_6 + v_7$,
$y_{1\times2} = v_4$,
$y_{1\times1} = v_7$,

(d) $v_j \geq 0$ (and integer if necessary), $j = 1,\ldots,7$.

The comparison of both model approaches shows clearly their different structure. The differences in the number of balance equations and variables here are negligible. This no longer applies to practical cases. Where an explicit model formulation is used the stepwise calculation seems to be advantageous. However, if the method of delayed pattern generation can be applied on the basis of the alternative calculation there is no need for the explicit model formulation. On the other hand, limiting the choice to "good" partitions (and corresponding cutting patterns) in the single-stage model may be often the only possible way to gain a hold on a problem. This particularly applies to two- and more-dimensional cutting processes with irregularly formed parts. For example, in the textile industry patterns ("cutting pictures") for cutting up parts of clothes from rolls are constructed by qualified specialists using their experience or with the help of a simulation procedure, or are taken from a previously stored stock of patterns. The single-stage model thus often results in good approximations to the optimal solution and stays manageable. The multi-stage model is ruled out here because the irregular order figures lead to the most varied intermediate figures of the remainder-pieces resulting from the one-cuts and being hardly identical with one another. The advantage of the multi-stage approach by means of one-cuts as simple partitions is only then effective when the figures are so regularly shaped that the same figure might be output for many different one-cuts. For more-dimensional cutting processes, this is possibly the case, when all the figures are rectangular blocks. With increasing dimensionality the size of both model approaches grows exponentially.

V. Concluding Remarks

Further differences in both model approaches can be seen in how they make allowance for certain objectives or restrictions of the cutting process which are neglected by the basic production-theoretical analysis of this chapter. The approach of alternative calculation shows itself to be very flexible (DYCKHOFF 1987). For example, impacts on costs which cannot be linearly combined with one-cuts, can be taken directly into account in costs-coefficients for the partitions. The same applies to restrictions on the construction of feasible partitions by reason of technical or organizational conditions. (For a more detailed discussion of the preconditions for the utilization of practical cutting models found in the literature cf. DYCKHOFF 1987.)

Interesting from the point of view of cost theory is an extension of the (efficiency) dominance concept in the direction of cost dominance of partitions, which goes back to MEERENDONK et al. (1963; cf. HOEK et al. 1983 and DYCKHOFF 1987). It has no complement in the cost-theoretical literature on converging production processes. On the whole, elastic joint production in conjunction with waste and excess production has been generally widely neglected (MÜLLER-MERBACH 1981) in the production- and cost-theoretical literature – in contrast to the literature of planning rigid joint production (cf. Chapter 9 of this book).

References

Andrews, G.E.: The Theory of Partitions, London et al. 1976.

Coffmann, E.G. jr., Garey, M.R., and Johnson, D.S.: Approximation algorithms for bin-packing – An updated survey, in: Ausiello, G., et al. (ed.): Approximation Algorithms for Computer System Design, Wien 1984, pp. 49 – 106.

Dantzig, G.B., and Wolfe, P.: Decomposition principle for linear programs, in: Operations Research, 1960, pp. 101 – 111.

Dyckhoff, H.: A new linear programming approach to the cutting stock problem, in: Operations Research, 1981, pp. 1092 – 1104.

Dyckhoff, H.: Elastische Kuppelproduktion geometrisch definierter Güter: Grundlagen wirtschaftlichen Zuschneidens und Packens, "Habilitationsschrift" (unpublished), FernUniversität Hagen 1987.

Fandel, G.: Produktion – Volume I: Produktions- und Kostentheorie, Berlin-Heidelberg-New York-Tokyo 1987.

Gardner, M.: Some packing problems that cannot be solved by sitting on the suitcase, in: Scientific American, 1979, Issue 10 (October).

Gardner, M.: Mathematische Spielereien, in: Spektrum der Wissenschaft, 1981, Issue 1 (January), pp. 6 – 11.

Gerhardt, C.: Bestimmungsmöglichkeiten optimaler Produktionsprogramme bei primärer Verbundproduktion, Ph.D.-Thesis, Hamburg 1966.

Gilmore, P.C., and Gomory, R.E.: A linear programming approach to the cutting stock problem, in: Operations Research, 1961, pp. 849 – 859.

Gilmore, P.C., and Gomory, R.E.: A linear programming approach to the cutting stock problem – part II, in: Operations Research, 1963, pp. 863 – 888.

Gilmore, P.C., and Gomory, R.E.: Multistage cutting stock problems of two and more dimensions, in: Operations Research, 1965, pp. 94 – 120.

Golden, B.L.: Approaches to the cutting stock problem, in: AIIE Transactions, 1976, pp. 265 – 274.

Heicken, K., and König, W.: Integration eines heuristisch-optimierenden Verfahrens zur Lösung eines eindimensionalen Verschnittproblems in einem EDV-gestützten Produktionsplanungs- und -steuerungssystem, in: Operations Research-Spektrum 1, 1980, pp. 251 – 259.

Hildenbrand, W.: Mathematische Grundlagen zur nichtlinearen Aktivitätsanalyse, in: Unternehmensforschung 10, 1966, pp. 65 – 80.

Hinxmann, A.I.: The trim loss and assortment problems: A survey, in: European Journal of Operational Research 5, 1980, pp. 8 – 18.

Hoek, G. van der, van den Meerendonk, H.W., and Wijmenga, R.T.: The elimination of convexly dominated cutting patterns in trimloss problems, Report 8333/0, Erasmus Universiteit, Rotterdam 1984.

Knolmayer, G.: Programmierungsmodelle für die Produktionsprogrammplanung – Ein Beitrag zur Methodologie der Modellkonstruktion, Basel-Boston-Stuttgart 1980.

Koopmans, T.C.: Analysis of production as an efficient combination of activities, in: Koopmans, T.C. (ed.): Activity Analysis of Production and Allocation, Cowles Foundation Monograph 13, New Haven 1951, pp. 33 – 97.

Meerendonk, H.W. van den, Kerbosch, J.A.G.M., Medema, P., and Schouten, J.H.: Some computational aspects of a trimloss problem, in: Statistica Neerlandica 17, 1963, pp. 37 – 47.

Müller-Merbach, H.: Die Konstruktion von Input-Output-Modellen, in: Bergner, H. (ed.): Planung und Rechnungswesen in der Betriebswirtschaftslehre, Berlin 1981, pp. 19 – 113.

Rao, M.R.: On the cutting stock problem, in: Journal of the Computer Society of India 7, 1976, pp. 35 – 39.

Stackelberg, H. v.: Grundlagen einer reinen Kostentheorie, in: Zeitschrift für Nationalökonomie, 1932, pp. 333 – 367, 552 – 590.

Stadtler, H.: A comparison of two optimization procedures for 1- and 1 1/2-dimensional cutting stock problems, in: Operations Research-Spektrum, 1988, pp. 97 – 111.

Wittmann, W.: Produktionstheorie, Berlin-Heidelberg-New York 1968.

Chapter 11

Trim Loss and Inventory Planning in a Small Textile Firm

HARALD DYCKHOFF and HERMANN GEHRING

I. Outline of the Problem

A factory obtains synthetic cloth wound onto rolls with a constant width of 366 cm but with varying lengths. The customers demand cloth with a narrower width than the initial width. The finished product is made by cutting whole rolls according to the parameters length and width of the customer's orders. A roll is cut on a cutting-machine with a single cutter into two part-rolls of the same length but with narrower width; the desired length is then obtained by rewinding (cf. Fig. 11.1).

The cutting process results in leftover rolls (cut pieces from whole rolls) which often can be used in filling of future orders. Very narrow rolls, however, can hardly ever be sold and have to be written off at the end of the year. Using customer's orders and sales forecasts as a starting point, the planning problem consists in organizing the cutting process such that the contribution margin is maximized and the residual stook at the end of the planning period does not exceed a specified area limitation. "Contribution margin" is to be understood as the sum of the net returns from the sale of rolls and the change in value of the stored rolls during the planning period.

The production flow can be described as a successive carrying out of customer's orders. As an order comes in, the stock-controller chooses a new or an already cut roll with which it is to be carried out. In choosing a roll he has to take into account the expected saleability of residual pieces. The fact that for a particular period of time (planning period) the expected orders are explicitly known is usually not considered.

Management expects a considerable increase in sales in the future, which would complicate the inventory problem and, as a consequence, overtax the stock-controller. In the course of a pilot study already performed in the past a model was developed which simultaneously considers all orders within the planning period (GEHRING et al. 1979). For the purposes of this chapter the model will be called Model A. Model A uses a selection of intuitively fixed cutting patterns, i.e. cutting patterns which appear to the stock-controller, by reason of his past experience, to be those least likely to lead to the production of unsaleable residual rolls. This, however, limits the set of solutions to mainly suboptimal solutions.

A further approach which has been developed (cf. DYCKHOFF 1981 and Chapter 10, Section IV) leads to Model B. This model uses special cutting pat-

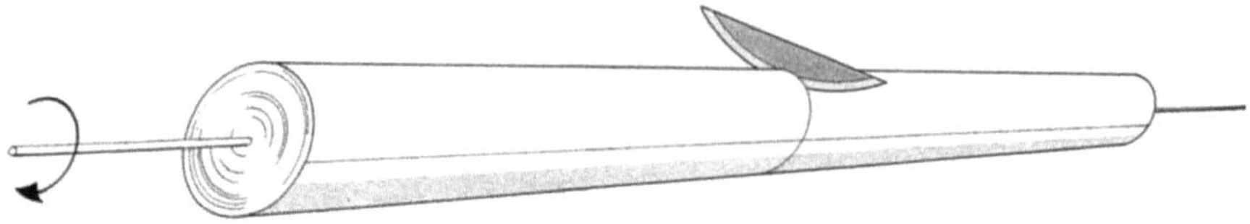

Fig. 11.1. One-cut technology

terns, so-called "one-cuts". Through dynamic application of one-cuts it is possible to generate all possible (effective) cutting patterns. Hence, by the aid of a suitable method of linear programming an optimal solution can be drawn from Model B.

The following sections contain a description of both models and a comparison using a concrete example. (The essential results of this study are originally published in German by DYCKHOFF and GEHRING (1982) and are partly updated here.)

II. Model A

At the beginning of the planning period the initial stock on hand and the customer's orders as regards length, width and price are known. To simplify the problem, rolls of the same width are joined together resulting in a total length. Orders are treated in an analogous manner, insofar as they correspond in width and price. Thus, the actual two-dimensional cutting problem is reduced to an one-dimensional problem as regards the width of synthetic cloth. This simplification is justifiable because customers will usually accept delivery of several rolls of the same width in fulfilment of a single order. Problems of this type are known as "1.5-dimensional" (cf. Chapter 12).

Let S be the set of widths of the initial stock on hand ("start-widths"), with the length l_j for $j \in S$, and let D be the set of widths for which orders exist ("order-widths"). Further, let K_j denote the set of orders k with order-width $j \in D$. For all of the start-widths $j \in S$ a selection of effective cutting patterns M_j is chosen. A cutting pattern is regarded as effective when it allows a roll to be cut in such a way as to leave at most one unsaleable (sc. in the current planning period) left-over roll. These cutting patterns define the set E of potential roll-widths at the end of the planning period ("end-widths"). Hence, $S \cup D \subset E$ applies.

The following equations guarantee the complete cutting of the initial stock on hand in the different specified cutting patterns, whereby x_{jm} indicates the length in which the width j is cut according to the pattern m (the degenerate cutting pattern "Do not cut" is always permissible):

$$(1) \qquad \sum_{m \in M_j} x_{jm} = l_j \quad (j \in S) \ .$$

Part rolls which come into being through cutting are either sold or become part of residual stock:

(2) $$\sum_{i\in S:\ b_i\geq b_j}\ \sum_{m\in M_i} h_{imj}x_{im} = \sum_{k\in K_j} z_{jk}+u_j \quad (j\in E)\ .$$

Here b_j is the width in units of length of the roll j, h_{imj} the frequency with which the end-width j is contained in the cutting pattern m for the start-width i, u_j the residual stock of the width j and z_{jk} the sold length in the width j for the order k.

The sold lengths z_{jk} are limited by means of minimum delivery quotas and upper sales limits:

(3) $$\underline{z}_{jk} \leq z_{jk} \leq \bar{z}_{jk} \quad (j\in D,\ k\in K_j)\ .$$

The stock on hand at the end of the planning period is restricted by an upper area limit:

(4) $$\sum_{j\in E} b_j u_j \leq f\ .$$

Non-negativity conditions exist as follows:

(5) $$u_j\geq 0 \quad (j\in E)\ , \qquad x_{jm}\geq 0 \quad (j\in S,\ m\in M_j)\ .$$

Finally, the sum of the net returns and the change in stock value is maximized:

(6) $$\max \sum_{j\in D}\ \sum_{k\in K_j} p_{jk}\, b_j\, z_{jk} + \sum_{j\in E} q_j\, b_j(u_j - l_j)\ .$$

The quantity p_{jk} indicates the net return per unit of area of the order k with width j; q_j indicates the value of stock per unit of area with width j, a value in which future saleability is shown. q_j refers to a capital value as the sum of the discounted net returns for this width; the value has to be estimated for the various widths on the basis of sales forecasts and past experience.

In principle every conceivable cutting pattern can be represented using Model A; however, an enumeration of all cutting patterns is only acceptable with smaller problems. Every additional cutting pattern increases in general the number of restrictions and variables. The number of variables even with average practical problems can go into the millions (cf. Chapter 10, Section III.2). If on the other hand only a small selection of relevant cutting patterns is used, chosen intuitively by an expert on the basis of his experience, then the number of restrictions and variables can be kept within reasonable limits; the disadvantage of this procedure, suggested by GEHRING et al. (1979), is that it leads to merely suboptimal solutions. For those who do not regard suboptimal solutions as being sufficient the question is raised how to determine optimal solutions.

GILMORE and GOMORY (1961, 1963) have developed a special method of solving the classic cutting problem, namely where the leftover rolls are of no value and are not subject to any restriction, a method which uses explicitly a relatively limited number of variables and still leads to optimal solutions. Similarly to the decomposition method developed by DANTZIG and WOLFE this approach considers at every pivot step of the revised simplex method auxiliary problems (knapsack problems) in order to generate a new improved cutting pattern which then is taken into the basis. The case under discussion is not a pure cutting problem in the classical sense. On the contrary, the final stock on hand is limited and has to be valued. These and further possible conditions require a modification of the original approach of GILMORE and GOMORY. (For recent developments concerning this subject see FARLEY 1988 and STADTLER 1988.) For the present study an analysis of these problems was dispensed with in favour of the development of a new model, viz. Model B.

III. Model B

Model B corresponds largely with Model A, although it is based on a different cutting technology. It conforms directly to the actual cutting, i.e. successive cutting with just one cutter. This results in special, simply structured cutting patterns, called "one-cuts". A relevant one-cut $[j;k]$ divides the width b_j into the part-widths b_k and $b_j - b_k$, whereby b_k is a customer-width. Every effective and complicated cutting pattern can be constructed successively using such one-cuts. For example, the one-cuts [9; 4], [5; 2], [3; 2] represent the following cutting pattern: the start-width 9 is to be divided into 1 piece with the customer-width 4, 2 pieces with the customer-width 2, and 1 piece with the end-width 1.

Model B can be represented in almost the same way as Model A if the sets E and M_j are now interpreted as follows: E denotes the set of all potential end-widths which arise as a result of relevant one-cuts; M_j denotes the set of all relevant one-cuts for a potential end-width $j \in E$. These sets can be determined quite simply by a recursive procedure.

With these definitions the formal notation of the restrictions (3), (4) and the objective function (6) can be taken over identically. The restrictions (1) and (2) are replaced by the following balance equations:

$$(7) \qquad l_j + \sum_{i \in E:\, b_i > b_j} \; \sum_{m \in M_i} h_{imj} x_{im} = \sum_{m \in M_j} x_{jm} + \sum_{k \in K_j} z_{jk} + u_j \qquad (j \in E) \; .$$

(Note: $l_j = 0$ for $j \notin S$; $\sum_{\varnothing} = 0$.) The restrictions (7) represent input-output relations for all potential end-widths. The input (left side of (7)) for the width j is composed of the initial stock l_j, if on hand, and new part-widths which result from cutting wider rolls. The output (right side of (7)) for width j consists of part-rolls of this width which are either cut further or sold or become part of the residual stock.

Thus, Model B consists of the objective function (6), the restrictions (3), (4),and (7) and the non-negativity conditions:

$$(8) \qquad u_j \geq 0 \quad (j \in E) \, , \qquad x_{jm} \geq 0 \quad (j \in E,\, m \in M_j) \; .$$

The set of feasible solutions of Model B contains an optimal solution because all relevant cutting combinations are considered. The size of the model cannot be specified in a generalized form; the size depends on the respective start-widths and customer's orders. If all effective cutting patterns were used in Model A both models would correspond as to the number of restrictions (in Model A the corresponding equations (1) and (2) can be combined analogous to (7) for $j \in S$). Model B, however, would contain considerably fewer variables (if sufficiently complex problems are considered).

IV. Comparison for a Concrete Example

Both models were tested by means of an example which was built using past data over a longer period. The example consists of 6 start-widths (Table 11.1) and 12 order-widths (Table 11.2). A larger sample would have been desirable for the comparison of the two models. Sales of synthetic cloth were still quite small at the time of the test so that it was not possible to set up a larger sample.

In Table 11.2 "U" denotes an upper sales limit and "L" a delivery obligation. Orders 3, 6, and 9 have to be delivered at price level 0, order 1 at cost price for new rolls (internal use or samples).

Table 11.1. Initial stock on hand

No.	Width [cm]	Length [m]
1 (new)	366	10800
2	160	1070
3	133	401
4	122	1310
5	91.5	1575
6	66	2834

Table 11.2. Customers' orders

No.	Width [cm]	Length [m]	L/U	Price [DM/m^2]
1	226	500	L	0.774
2	183	70	U	1.05
3	183	55	L	–
4	150	10747	U	1.05
5	140	2103	U	1.145
6	130	2168	L	–
7	122	1276	U	1.095
8	122	400	U	1.05
9	122	134	L	–
10	114	2168	U	1.10
11	105	250	U	1.15
12	91.5	3805	U	1.17
13	61	9207	U	1.–
14	55	1070	U	1.15
15	20	6166	U	1.10

L – lower limit, U – upper limit.

Proceeding from these data, three solutions were generated: the actual solution, a suboptimal solution using Model A and an optimal solution using Model B. The amount of the residual stock on hand of the actual solution was used as an upper limit for the two model solutions; widths of less than 6 cm, however, are regarded as trim loss and therefore not included in the final stock. The valuation factors q_j for the residual widths j in (6) were determined by means of a rule of thumb. This takes into account that the textile firm has defined easily marketable "standard widths" which are multiples of 1/6, 1/5, or 1/4 of the manufacturer's width of 366 cm (rounded off to 0.5 cm respectively). The factor q_j of a residual width j is calculated to be a fictitious earning per unit, discounted with a periodic rate of interest on the supposition that the residual width would be cut into standard widths which would be sold at the material costs and the trim loss would be kept as small as possible and written off as loss. Thus, $q_j = 0$ is valid especially for narrow widths j under 61 cm.

The actual solution is based on the real process in the firm during the considered time period. This solution is to be understood as the cutting patterns chosen by the stock-controller, the resulting deliveries and the residual stock on hand as well as the contribution margin which results through the valuation of sales and residual stock on hand using the objective function (6). Table 11.3 shows 14 one-cuts which represent the real cutting procedure.

The suboptimal solution derived from Model A is based on 50 intuitively selected, effective cutting patterns. Among them are cutting patterns which contain the 14 one-cuts of the actual solution. (The cutting patterns of the suboptimal solution are not represented here.)

Prior to deriving an optimal solution from Model B all relevant one-cuts and potential residual widths must be generated. An algorithm for this purpose is simply constructed and forms the basis of a matrix generator which effectively can be used in the application of Model B. The one-cuts of the optimal solution

Table 11.3. One-cuts of the actual solution

One-cut [cm]	Length [m]
[366; 183]	1957
[366; 150]	4572
[366; 140]	2103
[366; 130]	2168
[236; 114]	2168
[226; 150]	1603
[216; 150]	4572
[183; 91.5]	1832
[183; 61]	1957
[160; 55]	1070
[122; 61]	3625
[66; 20]	2834
[46; 20]	1666
[26; 20]	1666

Table 11.4. One-cuts of the optimal solution

One-cut [cm]	Length [cm]
[366; 226]	1283
[366; 150]	6729
[366; 114]	500
[366; 91.5]	620
[366; 61]	1668
[305; 61]	1668
[274.5; 91.5]	620
[252; 122]	500
[244; 114]	1668
[226; 150]	581
[226; 55]	202
[216; 150]	3437
[216; 20]	3292
[196; 61]	3079
[196; 55]	213
[183; 91.5]	495
[171; 55]	202
[160; 55]	250
[160; 20]	820
[141; 20]	213
[135; 61]	2377
[133; 20]	401
[121; 20]	213
[116; 55]	202
[113; 20]	401
[101; 20]	213
[93; 20]	401
[81; 61]	213

are listed in Table 11.4. Their interpretation will be demonstrated using the start-width 133 cm as an example: According to Table 11.1 the initial stock on hand for this width is 401 m. Table 11.4 shows that a strip of 20 cm width is cut from the whole length of this roll. From the remainder two further strips, each 20 cm wide, are successively cut. This results in a total of 1203 m with a width of 20 cm, with which order No. 15 (see Table 11.2) can partly be carried out. The remainder of the roll, now 73 cm wide, is not cut further (see Table 11.4) and becomes part of the residual stock on hand (see Table 11.5).

Table 11.5 shows the respective final stock, the total area of final stock and the contribution margin for each of the solutions. The total area of final stock is the same in each case, which means that for Model A and Model B it is at the upper limit.

A comparison of the three solutions shows that narrow widths (less than 61 cm) with poor sales chances are not found in the final stocks of the optimal solution. With respect to the total area the narrow width of 7.5 cm in the sub-optimal solution is of no consequence, when compared with the smallest widths of 46 cm and 6 cm found in the actual solution. Customer's orders are filled completely in all three cases, i.e. the earnings are the same.

Table 11.5. Comparison of results

	Actual solution	Suboptimal solution (Model A)	Optimal solution (Model B)
Width [cm]	Length [m]	Length [m]	Length [m]
135	–	–	702
133	401	–	–
122	–	1650	–
105	820	–	–
91.5	1434	–	–
76	1603	1489	581
74	–	–	2377
73	–	–	401
66	4572	6637	6271
46	1168	–	–
7.5	–	706	–
6	1666	–	–
Residual stocks [m²]	7580	7580	7580
Contribution margin [DM]	7380	7872	7930

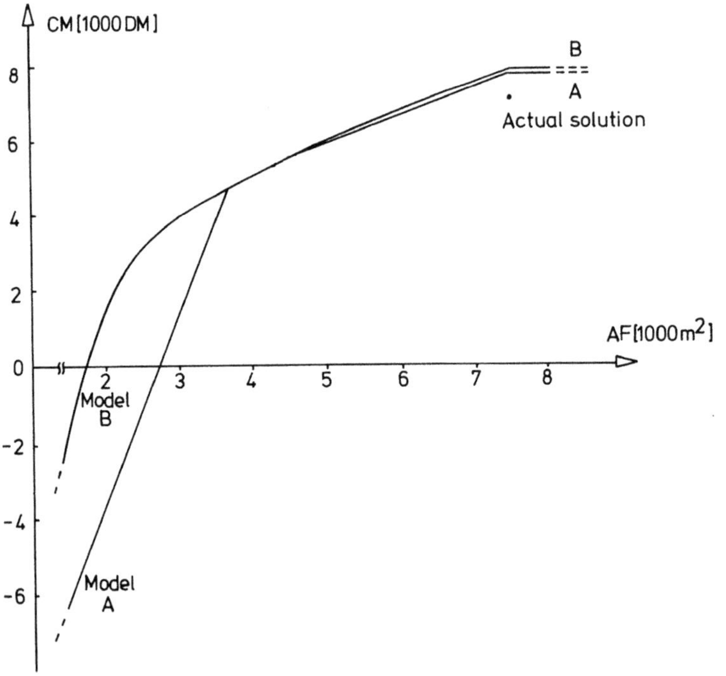

Fig. 11.2. Dependency of the contribution margin on the area of the residual stock on hand

Therefore, the differences in the contribution margins result from the differences in the structure of the final stocks. The relatively minor increase in the contribution margins of the two models in comparison with the actual solution (about 6.7% for Model A and 7.5% for Model B) is caused by the data in this example. The width of several orders (particularly internal orders and samples) was determined after other, preceding orders had been processed: In other words, these orders were not independent of stocks, so that an actual cutting problem for these widths did not exist.

If the (maximal) area of final stocks, AF, is varied within a parametric analysis, the trend of the contribution margin, CM, for Models A and B behaves as shown in Fig. 11.2. In both cases three areas with differing increases in contribution margins can be distinguished. The right-hand area shows that the contribution margin in Model A is constant and only increases minimally in Model B as a result of more advantageous cutting patterns with increasing areas of residual stocks. All customer's orders are carried out, those at price 0 only with their minimum lengths. If the area of the final stock is more limited ($AF \leq 7580$), orders at price 0 are carried out even in larger quantities, simply to keep to the reduction in area. In contrast to the left-hand area ($AF \leq 3900$), all orders with positive price are carried out up to their upper sales limit.

Because of the special selection of the 50 cutting patterns for Model A (which contain the actual solution) the curves in the right-hand and middle areas are practically the same for both models. This does not hold for the left-hand area. It can be seen quite clearly how big the differences between the optimal solution of Model B and the suboptimal solution of Model A can be in case of cutting problems with higher complexity.

Although Model B always generates an optimal solution, it can be more efficient, with regard to the required calculation time, to be content with suboptimal solutions. Using APEX III on a CYBER 175 the total calculation time was 0.2 CPU seconds for Model A (50 cutting patterns only) and 12 CPU seconds for Model B. For the above example Model B contains 335 restrictions and 2013 variables (matrix density ca. 0.9%). The size of Model A is fully dependent on the selection and number of cutting patterns; the 50 cutting patterns selected here lead to 39 restrictions and 101 variables. When all effective patterns are taken into account, Model A contains as many restrictions as Model B but in general many more variables. In connection with the (modified) Gilmore/Gomory-approach, however, Model A seems to represent a noninferior solution procedure, particularly with respect to the required calculation time (STADTLER 1988).

Because it uses one-cuts Model B on the other hand represents the actual cutting procedure more realistically. This is of great importance with respect to the acceptance of a planning model by the user.

References

Dyckhoff, H.: A new linear programming approach to the cutting stock problem, in: Operations Research, 1981, pp. 1092 – 1104.

Dyckhoff, H., and Gehring, H.: Vergleich zweier Modelle zur Lösung eines konkreten Verschnitt- und Lagerbestandsplanungsproblems, in: Operations Research-Spektrum 3, 1982, pp. 193 – 198.

Farley, A.A.: Practical adaptations of the Gilmore-Gomory approach to cutting stock problems, in: Operations Research-Spektrum, 1988, pp. 113 – 123.

Gehring, H., Gal, T., and Rödder, W.: Produktions- und Lagerbestandsplanung mit einem mehrstufigen Produktionsmodell, in: Gaede, K.W., et al. (eds.): Proceedings in Operations Research 8, Würzburg-Wien, 1979, pp. 390 – 396.

Gilmore, P.C., and Gomory, R.E.: A linear programming approach to the cutting stock problem, in: Operations Research, 1961, pp. 849 – 859.

Gilmore, P.C., and Gomory, R.E.: A linear programming approach to the cutting stock problem – part II, in: Operations Research, 1963, pp. 863 – 888.

Stadtler, H.: A comparison of two optimization procedures for 1- and $1\frac{1}{2}$-dimensional cutting stock problems, in: Operations Research-Spektrum, 1988, pp. 97 – 111.

Chapter 12

Classification of Real World Trim Loss Problems

HARALD DYCKHOFF, HERMANN-JOSEF KRUSE, DIETER ABEL, and TOMAS GAL

I. Introduction

Trim loss problems, also called cutting stock or depletion problems, are essential for production planning in some industries (e.g. in the paper, wood, metal, glass, plastic, and textile industries). In these industries the minimization of production costs is often approached in the following way: The raw material used is first produced in large standard sizes, possibly stored, and only later reduced to smaller sizes for in-plant processing or to meet customers' orders. This production pattern, on the one hand, involves a temporal uncoupling of raw material production and manufacturing of final products, and on the other hand it avoids a frequent resetting of production facilities which would be necessary if the product size required were produced from the input material in one step only. There are drawbacks, however, since an additional stage of production (the cutting operation) is required which generates useless remainder (trim loss). "Trim loss planning" is concerned with minimizing the negative effects thus created on the production costs.

In the Anglo-American as well as German scientific literature trim loss problems have not been dealt with before the mid fifties. Ever since, however, a variety of contributions has been published and is still being published. The papers by GOLDEN (1976), HINXMAN (1980), and MADSEN (1983) as well as a former German version of this paper (DYCKHOFF et al. 1984) which is extended and updated here give certain surveys of these contributions. It is necessary to point out that trim loss problems are deeply related to (bin) packing and (vehicle, pallet, or container) loading problems (cf. BROWN 1971); in some way they are two sides of the same coin. Therefore, also of eminent importance are surveys on approximation algorithms for bin packing by COFFMAN et al. (1984) and RAYWARD-SMITH and SHING (1983), on rectangular parts-nesting heuristics by ISRANI and SANDERS (1985) and RODE and ROSENBERG (1987) as well as on solution methods for two- and three-dimensional pallet or container loading problems by DOWSLAND (1985). The conclusion drawn by HINXMAN (1980) with respect to trim loss problems is:

"The state of the art is such that anyone faced with a trim-loss problem is likely to have to develop his own method, unless the Gilmore and Gomory approach is feasible. Such methods are almost bound to be heuristic. Study of papers, if any exist, on similar problems may prove helpful. However, heuristic methods tend to be strongly based in the particular problems to which they are applied, and apparently similar problems may turn out to require radically different heuristic methods".

Hence, in order to be able to make better use of the solution methods developed so far, it is often the case that people confronted with a concrete problem may reasonably try to find comparable problems in the literature. In particular, real world problems are of interest. For, not only the pure solution method is of importance; often, it is much more important to learn from case studies how a problem is approached, what assumptions and what kind of simplifications are made, which solution methods are discussed or even applied, and what experiences are gained. In short, the study of the framework which enables the successful (or perhaps inefficient) application of certain solution methods is, as a rule, at least as valuable as the knowledge of the methods themselves.

In this chapter, we give a survey of case studies for trim loss problems to be found in the German- and English-language literature. Based on these case studies we have developed a systematic, detailed catalogue of criteria which characterize trim loss (and also other related) problems in a differentiated manner. This classification system is problem-oriented and not method-oriented so that it allows for better comparisons of the individual problems studied in the literature. Since the planning situations underlying the trim loss problems considered will become better comparable and a translation of solution ideas is facilitated, this contribution complements the above mentioned method-oriented surveys. Furthermore, the case studies surveyed here are of interest not only as trim loss problems, but also as real world problems themselves. In one of these case studies dealing with shearing of steel bars, EILON (1960) states that "the problems [discussed in his paper] are characteristic of many other industrial operations, such as cutting paper or cloth, filling bottles or packets, portioning time to assignments of known duration, etc., and the results obtained here may, therefore, be applicable to many other fields in industrial engineering".

The present chapter is organized as follows: In Section II, the choice of the papers taken from the literature will be discussed. Section III explains the classification criteria proposed. In Section IV, the chosen case studies are characterized by these criteria and displayed in a "classification schedule".

II. Real World Trim Loss Problems in the Literature

The main body of scientific literature on trim loss problems can roughly be divided into two categories. The one – and by far the smaller – deals with case studies which are based on real trim loss problems. The other – by far the larger – is concerned with the development or improvement of models and solution techniques, taking abstractly formulated or already published real trim loss problems as a basis. For this study only the sources of the first category are consulted. They are characterized by the fact that the trim loss problems dealt with exist in practice and are described in sufficient detail regarding their structures and qualities relevant to the solution of the problem. In the German- and English-language literature available, we have thus identified 38 real trim loss problems dealt with in 36 contributions. (In a few cases, e.g. TROSSMANN (1983), there are described more than one case study which all have the same features with respect to our classification system.) The problems do not always fall definitely under one of the two categories. In such border-line cases they have been assigned arbitrarily.

The sources of the second category are not included in our analysis because they are not relevant to the first step done here. For further investigation they are clearly of importance, provided that they develop better models or solution methods for the trim loss problems described in the first category (e.g. COVERDALE and WHARTON (1976) and Chapter 11 of this book). For this purpose it will be helpful to consult the method-oriented surveys, mentioned in the introduction, as complementary surveys.

III. Classification Criteria

The goal to make trim loss problems comparable with each other can only be achieved by setting up a suitable catalogue of criteria. In order to select and systematize relevant classification criteria, a structural analysis of the real trim loss problems and of the corresponding models must be made. As regards the classification problem to be examined here, it suffices to deal with the steps of problem formulation, problem analysis, and particularly model building made in the model analysis of the case studies at issue.

For a rough classification of trim loss problems it is sufficient to take the step "model building" as the basis of systematization, by examining the individual case studies with respect to the model premises, determination and/or quantification of the model sizes, and the definition of the objective functions and the constraints. In this study, however, questions also have to be answered in the preceding steps of the model analysis, e.g. questions concerning the dimensions relevant to the trim loss problem considered and the planning situation in which the trim loss problem is embedded. They are indispensable for developing classification criteria.

The main criteria for a systematization, further differentiated in most cases, are:

- dimension,
- planning situation,
- goals,
- restrictions,
- solution approach.

1. Dimension

Since trim loss problems are usually concerned with the division of material, the objects marked down for the cutting operation are basically three-dimensional; but generally only one or two dimensions are relevant to the problem. Consequently, we can differentiate (see Fig. 12.1; cf. EVERSHEIM and HEMGESBERG 1977):

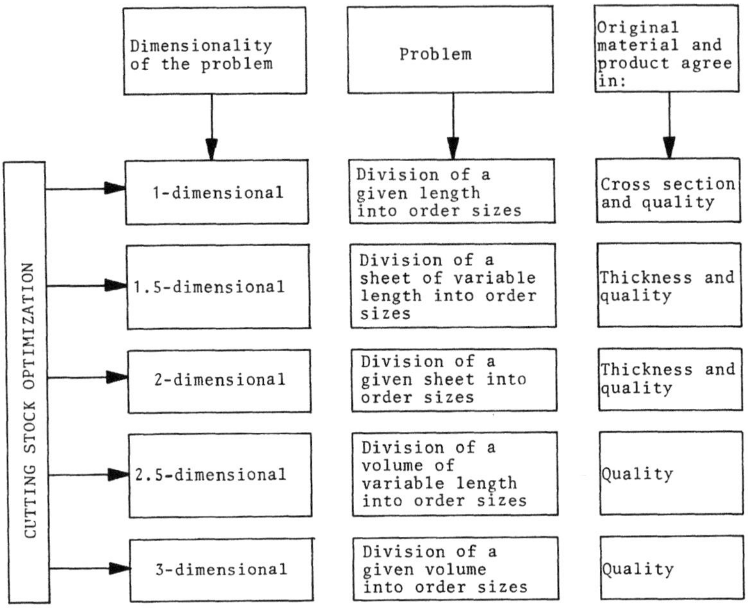

Fig. 12.1. Dimensionality of trim loss problems

- one-dimensional,
- two-dimensional, and
- three-dimensional trim loss problems.

Typical examples for one-dimensional problems are to be found in such lines of production where stocks of bars or rolls have to be cut into smaller pieces of the same cross section. Here only one dimension of the material (the "length") is relevant to the solution of the trim loss problem, whereas the cross section is fixed and hence does not affect the solution of the problem. Besides, there exist one-dimensional problems in such situations where the object to be divided has only one dimension anyhow, as for example when dividing spaces of time into smaller intervals of time (e.g. BROWN 1969). Such abstract "trim loss problems" are not considered in this chapter.

Two-dimensional problems exist in situations where flat material (e.g. metal sheets, chipboard, panes of glass, textiles) have to be divided into products of smaller square measures, but of the same, and thus not relevant, thickness of material.

Three-dimensional trim loss problems appear seldom, because in many real production situations one or two dimensions of the original material can be standardized, so that for each standardized type of material a problem with reduced dimensionality is obtained. Three-dimensional problems occur, however, in the form of so-called loading problems which are defined as "... the allocation of given items with known magnitude to boxes with constrained capacity, so as to minimize the number of boxes required ..." (EILON and CHRISTOFIDES 1971). Such problems are studied by MERLE and GRÜTZ (1978), STRASSBURGER (1984), and BAIR (1987); they may be looked at as abstract or generalized trim loss problems.

A further subdivision can be made when in addition to the existing degrees of freedom in generating the cutting patterns the lengths of individual pieces with respect to one relevant dimension are not uniquely determined, but are variable. A typical example for this is the cutting of rolls of a given width and total length, when it is up to the producer from how many rolls he wishes to cut the total length required. Such problems with two relevant dimensions of which one is fixed and the other variable in this sense are often designated as 1.5-dimensional (cf. HINXMAN 1980; there are other notions of 1.5-dimensionality in the literature, too). Extending this idea consistently, the existence of 2.5-dimensional trim loss problems is imaginable in such a way that two dimensions are fixed and the third is variable. In essence, 1.5- and 2.5-dimensionality refer to 1- and 2-dimensional problems, respectively, without integer requirements on the frequency of cutting patterns.

2. Planning Situation

In a real decision situation production as a whole consists of one optimization complex in which the cutting stock optimization is embedded. Since there are a lot of interdependencies between the various areas of production planning, in the sense of a total optimum to be aimed at, it is necessary to define the planning situation in such a way that as few as possible of the interdependencies are lost. Cutting stock optimization can mainly occur in connection with problems of sequencing and scheduling, timing, and inventory planning, as well as in connection with problems of choice among different manufacturing methods (e.g. different machines, cf. Fig. 12.2). Note that a cutting pattern itself represents a

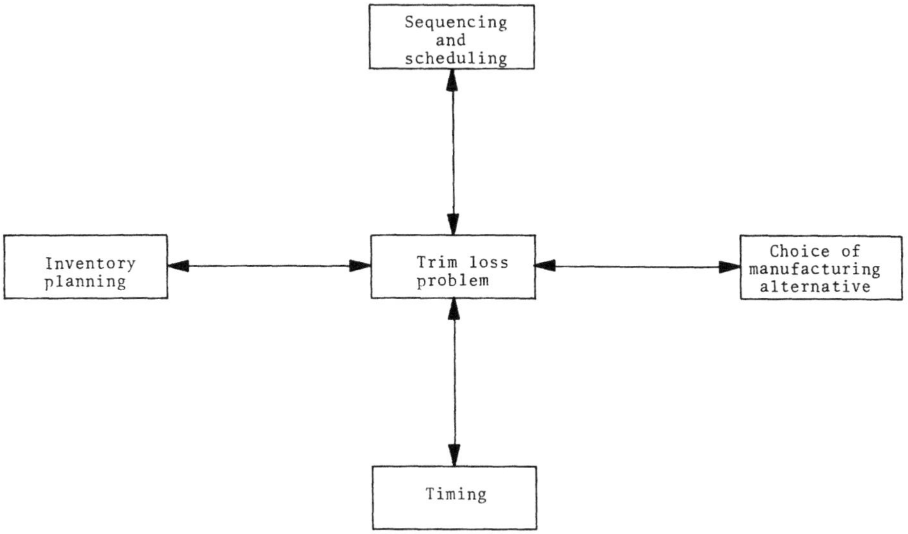

Fig. 12.2. Interdependencies between a trim loss problem and other problems of production planning

process of joint production with fixed yield coefficients (cf. Chapters 9 and 10 of this book).

Simultaneous planning is necessary to achieve a theoretical overall optimum. In most cases, however, this will lead to approaches which cannot be solved using the available solution techniques, or at least cause a prohibitive computational effort. There are at least two possibilities for the practical procedure:

- All areas are planned successively by treating one problem after the other. The solution of the preceding problem yields the constraints of the problem to be solved subsequently. The trim loss problem is thus considered in isolation.
- Provided that the bounds of a still acceptable effort in finding a solution are not exceeded, at least one or more problems can be integrated into a simultaneous planning approach.

3. Goals

Mapping of a trim loss problem into a decision model requires the selection of goal criteria and the formulation of objective functions. For the goodness of the model it is of fundamental importance that an objective function should map as genuinely as possible the concrete motives of action given in the real situation.

The design of an objective function to be used in the model firstly depends upon the main goal defined in advance by the organizational framework in which the planner works. Secondly, the formulation depends upon the section of reality considered; here one can differentiate for the time being between the two cases of isolated cutting stock optimization and simultaneous planning approaches.

In the case of isolated trim loss planning, the use of monetary goal variables such as costs, profit, or profit contribution can often be dispensed with. Goals in real terms are the minimization of the quantity of scrap (trim loss minimization) or of the original material used (input minimization). When the demand is exactly met, then both goals are equivalent in the sense that a solution implying minimal trim loss also implies simultaneously a minimal input, and vice versa. Both objectives are special cases of the goal "maximization of productivity" which can be expressed in the goal variable "relative trim loss" as the proportion of the total material used accounted for by scrap products. Such quantitative objectives, however, often involve difficulties. These objectives assign the same value per unit size to the original material used and the production surplus, irrespective of its structure and size, although input- and surplus output-pieces of different dimensions, but of the same weight, are not necessarily equally suited for a possible further processing. ("Production surplus" refers to quantities produced beyond demand, irrespective of whether they are suited for further use or are trim loss.) In such cases there are often different prices per unit size for original materials of different dimensions, so that cost minimization is a suitable goal. Moreover, provided that the products cut are valued differently according to unit size and that production does not need to satisfy demand, maximization of profit contribution can be taken into account.

Simultaneous planning approaches particularly imply the problem of combining individual goals connected with each of the planning areas into one single objective function, unless interactive approaches of multiobjective decision making are used. If the various individual goals can be measured in terms of monetary values the combined objective function corresponds to costs or profits, respectively. Sometimes this will meet with difficulties, particularly if timing is to be integrated into the simultaneous planning model. In such cases the overall goal can be called utility maximization. For the purpose of the present study this includes cases where different objectives have been discussed and implicitly taken into account of within heuristics. Case studies explicitly applying methods of multi-criteria decision making (MCDM) do not seem to exist.

4. Restrictions

A general framework for the classification of the restrictions affecting the trim loss problem is generated when the cutting operation is conceived as input/output process. A corresponding model describes the transition from input materials (production factors) to the desired output materials (products), as well as the technology determining the transition. These three elements of the model are describable in terms of quantity and value.

Therefore, from the quantitative properties of the model elements the following three categories of restrictions result:

- In a real planning situation input materials are available only in limited quantities and in certain sizes (input restrictions).
- The cutting technology cannot be varied at will, e.g. the amount of technology which is defined by the actually given parameters of the cutting machines is limited (cutting operation restrictions).
- Also the quantity of an output material which is to be produced, defined in terms of quality and sizes, is subject to certain restrictions; these are due to demand or existing internal orders (output restrictions).

Apart from these quantitative restrictions there may also be restrictions in terms of values. It is, for example, possible that restrictions in the form of defined cost budgets or minimal profit contributions limit the freedom of action. Our study of real trim loss problems, however, did not produce any example of

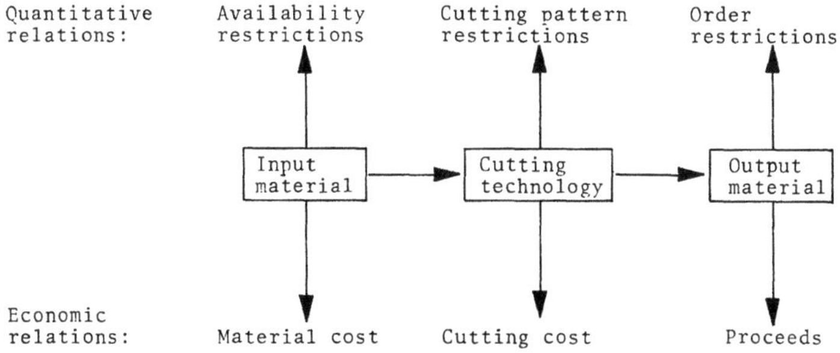

Fig. 12.3. Structure of the trim loss problem as input/output model

this kind, since in the contributions available cost and profit are taken into account only in the objective functions of the models. Therefore an investigation of restrictions in terms of values is dispensed with. For completeness, these components of value are, however, represented in Fig. 12.3, as well.

(a) Input Restrictions

Basically, every type of input material (equal size and equal quality) is available in finite quantities only, so that availability as a restriction should always be taken into account. Often, however, this restriction can be dispensed with, since a fresh supply of basic material can be produced or provided for within a short time, or great quantities of the basic material are, in any case, held in stock.

Input restrictions resulting from the "geometric properties" of the basic material are of greater importance for the construction of the model, since these restrictions influence the conditions of the cutting pattern generation. Here one can make the following distinction: Is all input material of the same size or must inputs of different sizes be taken into account in the model? If there are only pieces of the same size, then suitable cutting patterns have to be generated for this one size only, whereas in the other case they must be determined for each input size.

Depending on whether their sizes are known or not at the beginning of the planning period, inputs of different sizes can further be divided into deterministic and stochastic sizes. If the input sizes are determined at random one can furthermore distinguish whether the inputs are taken from stock or whether they are generated in a continuous flow of production. For example, rolls of plastic film taken from stock often have border defects of randomly fluctuating sizes. Another example is the continuous production of metal plates, slabs of precisely defined sizes are cut from ingots the sizes of which cannot be produced exactly.

As far as the deterministic inputs are concerned one distinguishes between the two cases in that either various standard sizes only or standard sizes and residual pieces of preceding cutting operations must be taken into account (cf. Fig. 12.4).

The above types of input restrictions do not necessarily lead to different model approaches. In many real cases simplifying premises are justified, so that the same model is applicable to various real problems. In the above example of plastic film with stochastically distributed border defects one can start from the premise that with a high level of significance the border defects do not exceed a certain depth of penetration into the film. The trim loss problem can be solved in a simplified way using a deterministic substitute problem, where a roll width is assumed which corresponds to the stocked width less the "maximal" depth of penetration of the border defects.

(b) Restrictions on the Cutting Operation

The conditions of the cutting operations exclusively concern the feasibility or suitability of cutting patterns; that is why they can also be designated as restric-

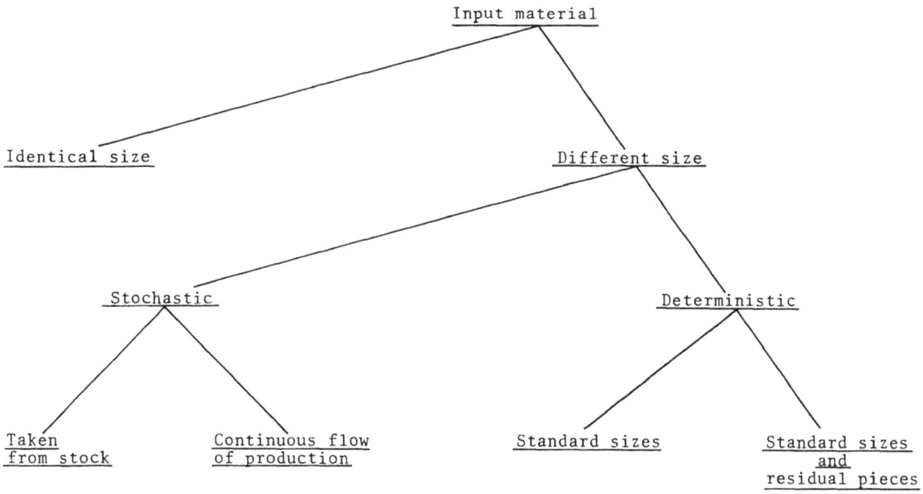

Fig. 12.4. Classification of the input material

tions on the cutting patterns. Restrictions of this kind can result from requirements with respect to organization or material, as well as from the cutting technology itself (cf. GÖTZ et al. 1972).

Restrictions due to organizational conditions mostly do not refer to the trim loss problem itself, but to the other areas of production planning mentioned in Section III.2. The requirement of carrying out customers' orders at a stretch may, for example, lead to the condition that cutting patterns should contain products of one customers' order only. Under such organizational conditions the interdependencies of other complexes of production planning are thus not established in the objective function of the model, but they limit the scope for generating cutting patterns.

Restrictions attributable to the material may also be subsumed under the input restrictions. Since, however, they mainly affect the trim loss problem in connection with the cutting technology, they are assigned to the restrictions on the cutting operation. They result from the physical properties or the structure of the material processed. Physical properties, for example, must be taken into account when glass is broken, where an unwanted jump of the break is to be avoided by keeping minimum distances between neighbouring cutting lines. Conditions of the structure of the material are, for example, of importance for the construction of cutting patterns when corrugated board or designed materials are processed.

Restrictions due to technological conditions represent the special properties of the cutting tools. First a distinction must be made between cutting operations yielding only rectangular products and those also yielding non-rectangular products. Non-rectangular material outlines are obtained when circular parts are punched or clothes are cut from lengths of dress material.

With regard to rectangular cutting operations a further distinction can be made between cuts that are made parallel to the edges (orthogonal) or made at

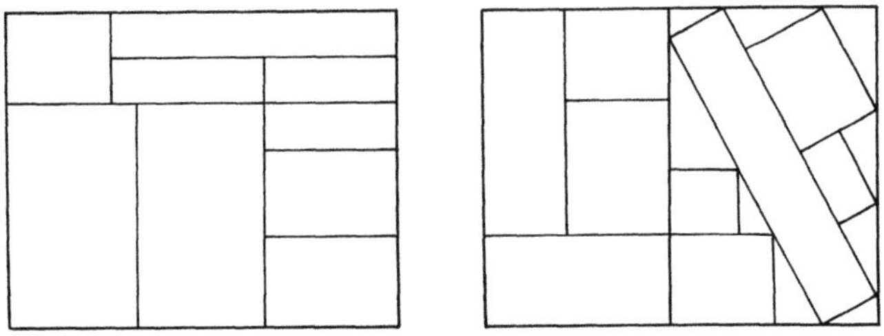

Fig. 12.5. Orthogonal and non-orthogonal cutting patterns

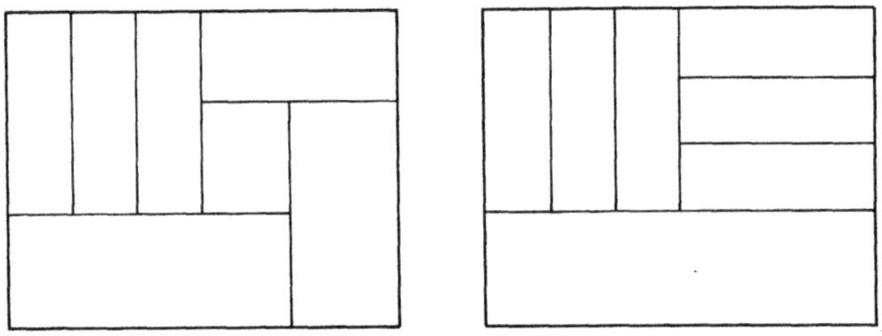

Fig. 12.6. Nested cutting pattern (left) and three-stage guillotine cutting pattern (right)

any angle to the latter (non-orthogonal; cf. Fig. 12.5). Since non-orthogonal cutting patterns have not been met with in any of the case studies investigated, a classification of these can be neglected.

Orthogonal cutting patterns can be divided into two further categories. If the cutting technology enables only straight cuts to be made from one edge of the material to the opposite one, as is the case when cutting paper with cutting machines or when breaking panes of glass, then we speak of guillotine cutting patterns. If, however, it is permitted to discontinue the cut at any place in the material, then interlocked (or nested) cutting patterns can also be cut. With regard to guillotine cutting it is reasonable to distinguish between such cutting patterns which may imply any number of parallel cuts and those whose number of parallel cuts is limited (limitations with respect to the number of cutting knives). By multi-stage cutting in the sense of a multi-stage production process the guillotine cuts can also yield more complicated cutting patterns (cf. Fig. 12.6).

(c) Output Restrictions

The type and quantity of the output produced must comply with the demand. In order-oriented production the demand is determined by customers' orders. The customers may generally be assumed to place orders of certain sizes which the producer must fulfil exactly. Nevertheless, a deviation is sometimes admissible (e.g. in the paper industry the allowed variation is $\pm 5\%$ in normal practice). Even in the case of a strictly defined quantity ordered, the quantity produced can

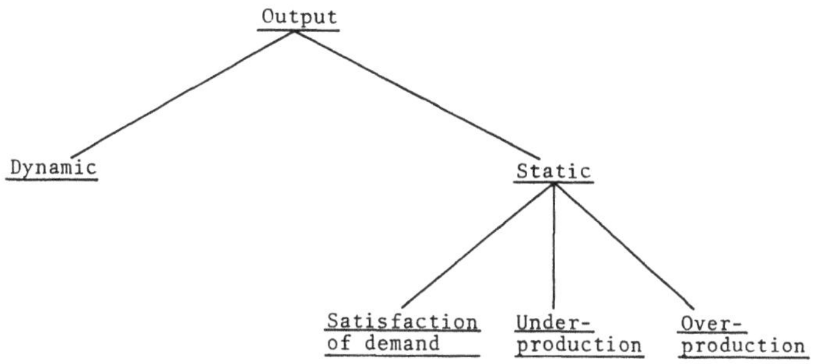

Fig. 12.7. Classification of the output

differ from the former, when the surplus quantities are transferred to stock and the goods produced in too little quantities are bought elsewhere or taken from stock.

In the case of a market-oriented production, however, the demand over a planning period is estimated by the sales department, the demand forecasted being passed on to the manufacturing division as an internal order. Naturally the estimates of the sales department will imply uncertainties. Consequently, a limited over- or underproduction should be allowed.

The above cases of exact satisfaction of demand, of over- or underproduction are mostly relevant in a static model only. In a dynamic planning situation the orders come in successively and every order coming to hand requires an immediate decision upon the input material to be used (cf. Fig. 12.7).

5. Solution Approaches

The classification criteria described above result from following the first three steps of the model-analytic procedure. Just as other classifications pursue a specific aim so does this classification, here in the sense that it will enable comparable problems already solved to be discovered more easily and thus solution approaches to be applied to other problems and new solution ideas to be obtained. Consequently, it might also be useful to list the solution approaches chosen – which represent a later step of the model analysis – in addition to the problem classification criteria already described.

The classical approach is characterized by a two-stage procedure. First, suitable cutting patterns are constructed by enumeration often using heuristics. Then the number of each cutting pattern to be realized is determined using linear programming; subsequently integrality conditions can be taken into account by rounding off. Since total enumeration will usually lead to an enormous number of potential cutting patterns and thus variables, even when dealing with medium problem sizes, a limited selection is made from the potential cutting patterns, as a rule. This heuristic procedure is avoided, if the approach of "delayed pattern generation" by GILMORE and GOMORY (1961, 1963, 1965) can be applied. Within the framework of the revised simplex method this approach yields the optimal cutting patterns implicitly by considering knapsack problems as auxiliary problems at each iteration step.

In order to take into account integer constraints on the (cutting pattern) variables, exact methods of discrete optimization might be applied instead of rounding strategies. This procedure, however, will lead to an acceptable computational effort only if small problem sizes, i.e. few potential cutting patterns in a small number of pieces, are dealt with; even then the computational effort is not justifiable unless the solution of the trim loss problem is the basis of large-scale or mass production. Apart from that, discrete optimization methods are sometimes also used for determining optimal sequences of cutting patterns. As a rule, however, heuristics will be used in these instances; they are framed especially for the individual decision situation and will ensure a "good" solution with relatively little effort (one of the earliest and often used is the heuristic of PIERCE 1964). Among the other solution approaches, simulation is worth mentioning. It will be applied in such situations in which one or several model parameters are subject to stochastic influences, or when the trim loss problem is complicated by irregularly shaped demand sizes, e.g. when cutting out parts of clothing.

IV. Classification Schedule

A survey of the structure of the 38 real trim loss problems analysed is given in the following classification schedule, assigning to the individual problems (columns of the schedule) their qualities with regard to the specific classification criteria (rows of the schedule). This schedule (see Table 12.1) thus also allows an immediate comparison of all problems investigated. The problems have been arranged chronologically; within the same year the arrangement is alphabetical. The order of the criteria is in line with the systematics developed in Section III. Some special features of the classification schedule should be pointed out:

- A row for 2.5- and 3-dimensional trim loss problems has not been included in the schedule, since problems of this kind do not occur in the literature analysed.
- Multiple labelling (marked by black circles) in a column with respect to one "classification" criterion means that the corresponding features occur alternatively (e.g. different solution methods) or in combination (e.g. planning situations). With circles denote a subordinated feature.
- With respect to the sources analysed it is not always possible to assign a problem to one row definitively. In such cases the most plausible labels (or in cases of doubt no labels) have been set up in type. This procedure is particularly related to the rows "limited availability" and "multi-stage cutting operation". The former row has been labelled only if the corresponding criterion is explicitly relevant to the model building process or the solution method. (Thus, in the rows "limited availability" and "identical size" there can be at most one label in all, if the rows "satisfaction of demand" or "overproduction" have been marked simultaneously.)

- The assignment of the solution methods has been made very roughly only, since – as mentioned before – that is beyond the primary purpose of this contribution. The row "Others" includes solution approaches such as simulation, branch and bound, or analytic and other methods. The marks in the row "GILMORE/GOMORY" are white circles (instead of black ones), since this approach has either been tested only and then rejected, or forms a part of a subsequent heuristic.

In the last column of the classification schedule we have entered the frequency with which the individual criteria will occur on the whole in the real trim loss problems described in the literature. We shall, however, dispense with a detailed statistical evaluation of interdependencies between the occurrence of criteria (e.g. by cluster analysis or by computing correlations). When the trim loss problems analysed are considered to be a random sample, then the number of 38 seems to be too small to make reliable statements. It is anyhow more realistic to assume that the majority of the published case studies stand vicariously for an (unknown) multitude of practical trim loss problems, i.e. to be a type of trim loss problem themselves. Forming clusters among the 38 problems investigated could at best lead to a typification of types, indicating nothing about their actual frequency in reality. For example, the one-dimensional, pure trim loss minimization problem without any other restrictions ("standard problem") which has been described in the literature as the first real problem (PAULL 1956) is certainly underrepresented. This is due to the fact that the approach of implicit (or "delayed") cutting pattern generation developed by GILMORE and GOMORY is an efficient solution technique for this type of problem, so that further publications of case studies dealing with this type are not necessary. It is for the same reason that the Gilmore- and Gomory-approach seldom occurs in the schedule, and if it does so, it is mentioned in connection with other solution procedures.

Although frequency of occurrence in reality cannot be concluded from frequency of occurrence in the classification schedule, it may nevertheless be stated that specific criteria are mentioned relatively infrequently. They include:

- simultaneous planning with timing and choice of manufacturing methods;
- the goals input minimization, maximization of productivity, or profit contribution;
- stochastic input variables;
- cutting patterns with non-rectangular output or with feasible interlocking;
- dynamic planning situations;
- "exact" solution techniques, such as the Gilmore- and Gomory-approach or integer programming solution methods, as well as "other" solution techniques, such as e.g. simulation.

The other criteria, however, are mentioned almost equally as often. As already pointed out, 2.5- and 3-dimensional problems are not to be found in the schedule at all. On the contrary, there are certainly features of problems that are not mentioned in the classification schedule. So, for example, the trim loss problem in STAHLKNECHT (1970) is complicated in that the input length is not given, but has to be optimally determined simultaneously, i.e. it is an assortment problem (cf. HINXMAN 1980).

Table 12.1. Classification schedule for real world trim loss problems

Classification criteria / Cutting of ... (Authors and year of publication)		Rolls of paper (PAULL 1956)	Steel coils (METZGER 1958)	Rolls of paper (FÖRSTNER 1959)	Steel bars (EILON 1960)	Steel bars (EILON 1960)	Punching of circles from rectangular plates (LAMPL / STAHL 1965)	Steel slabs (PELZER / RUTH 1966)	Chipboards (WIHL 1969)	Steel bars (STAHLKNECHT 1970)	Rolls of paper (HAESSLER 1971)	Glass planes for double glazing (GÖTZ et al. 1972)	Transformer sheet (GOSWAMI 1973)	Glass plates (DYSON / GREGORY 1974)	Steel slabs (TILANUS / GERHARDT 1976)	Steel coils (HAESSLER 1977)	Steel coils (HAESSLER 1977)
		1	2	3	4	5	6	7	8	9	10	11	12	13	14	15	16
Dimension	1 - dimensional	●	●		●	●		●		●	●				●		
	1.5 - dimensional			●									●			●	●
	2 - dimensional						●		●			●		●			
Planning situation	Cutting stock planning (isolated)	●	●			●	●		●	●							
	with sequencing and scheduling				●							●		●		●	●
	with inventory planning			●				●					●				
	with timing							●							●		
	with choice of manufacturing methods										●						
Goals	Trim loss minimization	●	●			●	●	●	●	○		●	●				
	Input minimization									●							
	Maximization of productivity						●										
	Cost minimization				●						●					●	●
	Maximization of profit contribution			●													
	Utility maximization													●	●		
RESTRICTIONS – Input	Limited availability			●		●	●	●				●	●		●	●	
	Identical size	●			●					●	●						●
	Diff. size – Determ. – Standard size			●			●		●					●		●	
	Diff. size – Determ. – Residual pieces											●	●				
	Diff. size – Stoch. – taken from stock		●			●											
	Diff. size – Stoch. – Continuous prod.							●							●		
RESTRICTIONS – Cutting operations	Material		●					●	●						●	●	●
	Organization			●	●					●	●	●		●	●	●	●
	Technology – Non-rectangular						●										
	Technology – Rectangular – Nested																
	Technology – Rectangular – Guillotine – Limited		●						●		●	●	●	●			
	Technology – Rectangular – Guillotine – Unlimited	●		●	●	●		●		●					●	●	●
	Technology – Multi-stage cutting oper.				●	●			●			●	●	●			
RESTRICTIONS – Output	Dynamic																
	Static – Satisfaction of demand	●	●		●	●	●					●			●		
	Static – Underproduction								●		●					●	●
	Static – Overproduction			●				●	●	●	●		●	●		●	●
Solution approaches	Classical approach	●	●	●	○	●	●		●	●							
	GILMORE / GOMORY													○			
	Heuristics									○	●	●		●		●	●
	Others				●		○	●					●		●		

Table 12.1. (continued)

Carpet rolls (LITTON 1977) 17	Steel plates (NIEDERHAUSEN 1977) 18	Steel reinforcement bars (STAINTON 1977) 19	Corrugated board (SCHEPENS 1978) 20	Fleece in the form of rolls (GEHRING et al. 1979) 21	Plastic film (HAESSLER 1979) 22	Steel slabs (HAESSLER et al. 1979) 23	Glass planes for double glazing (thermopanes) (MADSEN 1979) 24	Parts of trousers from lengths of material (ELLINGER et al. 1980) 25	Tubes for central heating radiator units (HEICKEN / KÖNIG 1980) 26	X-ray films (MALLERET 1980) 27	Glass planes for double glazing (ELLINGER et al. 1981) 28	Rolls of isolated tape (SCULLI 1981) 29	Steel blooms (TOKUYAMA / UENO 1981) 30	Steel plates (TOKUYAMA / UENO 1981) 31	Metal sheets (LAUBENSTEIN et al. 1982) 32	Pieces of clothes (TROBMANN 1983) 33	Worktops (ROBERTS 1984) 34	Steel coils (WÄSCHER et al. 1985) 35	Corrugated cardboard (MOLL 1985) 36	Rolls of paper (BARTMANN 1986) 37	Corrugated cardboard (BOOKBINDER / HIGGINSON 1986) 38	Frequencies of the criteria
●		●				●			●			●	●					●	●		●	17
			●	●	●			●												●		9
	●						●			●	●			●	●	●	●					12
	●										●											8
		●	●		●	●	●	●	●	●			●		●	●	●	●	●	●	●	21
●		●		●						●			●	●			●	●		●		12
						●												○		○		3(2)
			●							●		●	●									5
●	●						●				●											12(1)
									●													2
												●										2
			●		●			●		●						●						9
				●																		2
		●				●							●	●	●		●	●	●	●	●	12
●	●	●		●		●							●	●			●	●				17
					●		●		●	●	●					●			●			12
			●			●		●							●						●	10
●	●	●		●										●			●	●				9
												●	●									4
																				●		3
	●		●			●	●	●					●			●	●					14
●		●	●		●	●	●	●	●	●	●		●		●	●			●	●	●	25
	●							●								●						4
														●			●					2
			●	●	●				●	●					●			●	●		●	15
●		●				●	●				●	●	●							●		17
●		●	●	●	●		●			●	●			●	●			●		●		18
●																					●	2
	●	●					●	●			●		●	●			●					15
				●	●	●				●								●	●	○		10(1)
			●		●				●	●		●			●	●		●	●	○		18(1)
				●				●	●	○												10(2)
					○	○	○															(4)
		●	●		●	●	●			●	●		●	●	●	●	●	●	●	○	●	20(2)
●	●											●								●	○	8(2)

It should be noted that nearly half of the case studies analysed fall to the metal industry's share (17 from 38). On the second place ranges the paper and corrugated cardboard industry (7). The remaining trim loss problems are almost equally distributed among the other branches mentioned in the introduction: glass (4), textile (4), wood and plastics (5), and other (1).

The majority of the solution methods are heuristic ones. These include the classical approach because of the limited number of considered cutting patterns. It has been applied particularly before GILMORE and GOMORY (1961) developed their approach. The latter is an efficient solution technique which is applicable to several types of trim loss problems (cf. FARLEY 1988 and STADTLER 1988). As has also been stated by HINXMAN (1980), there is, however, a variety of problems which cannot be solved exactly or efficiently using the approaches available. This becomes apparent by the fact that especially in the more recent publications, real world trim loss problems are very often solved by heuristics. An even more important reason for this development is the fact that, in practice as a rule, there are several objectives that have to be taken into account when solving a trim loss problem. Besides (explicit) approaches of multi-criteria decision making, heuristics are often a flexible instrument for this task.

References

Bair, Th.: Eine Heuristik zur Lösung des dreidimensionalen Verschnittproblems, Diploma-thesis, FB Informatik, Universität Dortmund 1987.

Bartmann, D.: Entwicklung und Einsatz eines DV-Systems zur Auftragsverplanung bei einem Papierhersteller, in: Angewandte Informatik, 1986, pp. 524 – 532.

Bookbinder, J.H., and Higginson, J.K.: Customer service vs. trim waste in corrugated box manufacture, in: Journal of the Operational Research Society, 1986, pp. 1061 – 1071.

Brown, A.R.: Selling television time – An optimization problem, in: The Computer Journal, 1969, pp. 201 – 207.

Brown, A.R.: Optimum Packing and Depletion – The Computer in Space- and Resource-usage Problems, London 1971.

Coffman, E.G., Garey, M.R., and Johnson, D.S.: Approximation algorithms for bin-packing – An updated survey, in: Ausiello, G., et al. (eds.): Approximation Algorithms for Computer System Design, Wien 1984, pp. 49 – 106.

Coverdale, I., and Wharton, F.: An improved heuristic procedure for a nonlinear cutting stock problem, in: Management Science, 1976, pp. 78 – 86.

Dowsland, W.B.: Two and three dimensional packing problems and solution methods, in: New Zealand Operational Research, 1985, pp. 1 – 18.

Dyckhoff, H., Abel, D., Kruse, H.-J., and Gal, T.: Klassifizierung realer Verschnittprobleme, in: Zeitschrift für betriebwirtschaftliche Forschung, 1984, pp. 913 – 929.

Dyson, R.G., and Gregory, A.S.: The cutting stock problem in the flat glass industry, in: Operational Research Quarterly, 1974, pp. 41 – 53.

Eilon, S.: Optimizing the shearing of steel bars, in: Journal of Mechanical Engineering Science, 1960, pp. 129 – 142.

Eilon, S., and Christofides, N.: The loading problem, in: Management Science, 1971, pp. 259 – 268.

Ellinger, T., Asmussen, R., and Schirmer, A.: Materialeinsparung durch optimale Zuschnittplanung, RKW-No. 674, Eschborn 1980.

Ellinger, T., Asmussen, R., and Schirmer, A.: Rationalisierung durch Verschnittoptimierung, RKW-No. 716, Eschborn 1981.

Eversheim, W., and Hemgesberg, G.: Entwicklung eines Systems zur optimalen zweidimensionalen Verschnittoptimierung in der Einzel- und Kleinserienfertigung mit Hilfe dialogfähiger Rechenanlagen, Opladen 1977.
Farley, A.A.: Practical adaptations of the Gilmore-Gomory approach to cutting stock problems, in: Operations Research-Spektrum, 1988, pp. 113–123.
Förstner, K.: Zur Lösung von Entscheidungsaufgaben bei der Papierherstellung, in: Zeitschrift für Betriebswirtschaft, 1959, pp. 693–703 and pp. 756–765.
Gehring, H., Gal, T., and Rödder, W.: Produktions- und Lagerbestandsplanung mit einem mehrstufigen Produktionsmodell, in: Gaede, K.-W., et al. (eds.): Proceedings in Operations Research 8, 1979, pp. 390–396.
Gilmore, P.C., and Gomory, R.E.: A linear programming approach to the cutting-stock problem, in: Operations Research, 1961, pp. 849–859.
Gilmore, P.C., and Gomory, R.E.: A linear programming approach to the cutting-stock problem – part II, in: Operations Research, 1963, pp. 863–888.
Gilmore, P.C., and Gomory, R.E.: Multistage cutting stock problems of two and more dimensions, in: Operations Research, 1965, pp. 94–120.
Golden, B.L.: Approaches to the cutting stock problem, in: AIIE Transactions 8, 1976, pp. 265–274.
Goswami, P.: Wertanalysebeispiel, Reduktion des Blechverschnitts im Trafobau, in: Industrielle Organisation, 1973, pp. 343–347.
Götz, H., Littger, K., and zur Steege, R.: TRIM – ein Verfahren zur Behandlung des zweidimensionalen Verschnittproblems und sein Einsatz bei der Flachglas AG DELOG-DETAG, IBM-Form K 12-1044, Stuttgart 1972.
Haessler, R.W.: A heuristic programming solution to a non-linear cutting stock problem, in: Management Science, 1971, pp. B793–B802.
Haessler, R.W.: A procedure for solving the 1.5-dimensional coil slitting problem, in: AIIE Transactions 10, 1977, pp. 70–75.
Haessler, R.W.: Solving the two-stage cutting stock problem, in: Omega, 1979, pp. 145–151.
Haessler, R.W., and Vonderembse, M.A.: A procedure for solving the master slab cutting stock problem in the steel industry, in: AIIE Transactions 11, 1979, pp. 160–165.
Heicken, K., and König, W.: Integration eines heuristisch-optimierenden Verfahrens zur Lösung eines eindimensionalen Verschnittproblems in einem EDV-gestützten Produktionsplanungs- und -steuerungssystem, in: Operations Research-Spektrum 1, 1980, pp. 251–259.
Hinxman, A.I.: The trim-loss and assortment problems – A survey, in: European Journal of Operational Research 5, 1980, pp. 8–18.
Israni, S.S., and Sanders, J.L.: Performance testing of rectangular parts-nesting heuristics, in: International Journal of Production Research, 1985, pp. 437–456.
Lampl, T., and Stahl, J.: Über den optimalen Zuschnitt von Plattenmaterial, in: Unternehmensforschung, 1965, pp. 187–197.
Laubenstein, T., Schneeweiß, C., and Vaterrodt, H.J.: Verschnittoptimierung im praktischen Einsatz, in: Operations Research-Spektrum 4, 1982, pp. 229–236.
Litton, C.D.: A frequency approach to the one-dimensional cutting problem for carpet rolls, in: Operational Research Quarterly, 1977, pp. 927–938.
Madsen, O.B.G.: Glass cutting in a small firm, in: Mathematical Programming, 1979, pp. 85–90.
Madsen, O.B.G.: Computerized cutting and order-spread minimization procedures and their implementation on a microcomputer, IMSOR Research Report 18A, Lyngby (Denmark), 1983.
Malleret, V.: A two-step solution to a cutting and scheduling problem: a case study, Report prepared for EURO IV Congress, Cambridge (U.K.), 1980.
Merle, G., and Grütz, M.: Entwicklung eines heuristischen Verfahrens zur quantitativen Lösung des Verpackungsproblems im Versandhandel, in: Späth, H. (ed.): Fallstudien Operations Research, Volume 2, München-Wien 1978, pp. 108–130.
Metzger, R.W.: Elementary Mathematical Programming (Chap. 8: Stock slitting), New York 1958.
Moll, R.: Verfahren zur mehrzielorientierten Verschnittplanung in der Wellpappenindustrie, Bergisch Gladbach-Köln 1985.
Niederhausen, H.P.: Automatische Schnittplan-Optimierung, in: VDI-Berichte No. 277, 1977, pp. 121–126.

Paull, A.E.: Linear Programming: a key to optimum newsprint production, in: Pulp and Paper Magazine of Canada, 1956, pp. 146 – 150.
Pelzer, H., and Ruth, K.A.: Steuerung des Stoffflusses und optimale Brammenaufteilung durch Einsatz einer elektronischen Rechenanlage, in: Stahl und Eisen, 1966, pp. 100 – 106.
Pierce, J.F.: Some large-scale production scheduling problems in the paper industry, Englewood Cliffs 1964.
Rayward-Smith, V.J., and Shing, M.T.: Bin packing, in: Bulletin of the IMA, 1983, pp. 142 – 146.
Roberts, S.A.: Application of heuristic techniques to the cutting stock problem for worktops, in: Journal of the Operational Research Society, 1984, pp. 369 – 377.
Rode, M., and Rosenberg, O.: An analysis of heuristic trim loss algorithms, in: Engineering Cost and Production Economics 12, 1987, pp. 71 – 78.
Schepens, G.: The development and implementation of an interactive cutting model for the corrugated board industry, in: Hax, A.C. (ed.): Studies in Operations Management, Amsterdam-New York-Oxford 1978, pp. 286 – 301.
Sculli, D.: A stochastic cutting stock procedure – Cutting rolls of insulating tape, in: Management Science, 1981, pp. 946 – 952.
Stadtler, H.: A comparison of two optimization procedures for 1- and $1\frac{1}{2}$-dimensional cutting stock problems, in: Operations Research-Spektrum, 1988, pp. 97 – 111.
Stahlknecht, P.: Operations Research, Braunschweig 1970, pp. 94 – 104.
Stainton, R.S.: The cutting stock problem for the stockholder of steel reinforcement bars, in: Operational Research Quarterly, 1977, pp. 139 – 149.
Straßburger, F.X.: Die optimale Beladung von Containern: Lösungsmöglichkeiten zum dreidimensionalen Verschnittproblem, Diploma-thesis, FB Wirtschaftswissenschaften, Universität München 1984.
Tilanus, C.B., and Gerhardt, C.: An application of cutting stock in the steel industry, in: Haley, K.B. (ed.): Operational Research '75, Amsterdam et al. 1976, pp. 669 – 675.
Tokuyama, H., and Ueno, N.: The cutting stock problem in the iron and steel industries, in: Brans, J.P. (ed.), Operational Research '81, Amsterdam et al. 1981, pp. 809 – 823.
Tokuyama, H., and Ueno, N.: The cutting stock problem for large sections in iron and steel industries, in: European Journal of Operational Research 22, 1985, pp. 280 – 292 (see also Tokuyama/Ueno 1981).
Troßmann, E.: Verschnittoptimierung – dargestellt an Beispielen aus der Textilindustrie, Berlin 1983.
Wäscher, G., Carow, P., and Müller, H.: Entwicklung eines flexiblen Verfahrens für Zuschneideprobleme in einem Kaltwalzwerk, in: Zeitschrift für Operations Research, 1985, pp. B209 – B230.
Wihl, W.: Verschnittminimierung beim Zuschnitt aus Platten in Standardgrößen, in: Bürotechnik und Automation, 1969, pp. 186 – 192.

Chapter 13

Standard Software for Cutting Stock Management

HARALD DYCKHOFF, UTE FINKE, and HERMANN-JOSEF KRUSE

I. Introduction

Cutting stock processes cause a series of problems for which computer support is of importance, if not indispensable. Along with the questions of material management, especially the question of how much raw material can be saved by optimizing the cutting process and thus reducing waste, other aspects of computer-based production planning and control are touched on, including the embedding of the cutting process in a comprehensive CIM system.

Usually only large companies which have the necessary development capacity available can afford to produce their own tailor-made software. Most of the companies concerned, however, are small and medium-sized firms which could not possibly develop their own programs and are therefore dependent on the services of hardware manufacturers, software producers, or other companies offering suitable programs. In addition to the external development of an individual program the (usually cheaper) alternative is to buy standard software where it exists, and to modify it where necessary.

The demand thus caused for standard software for cutting stock problems is faced with a supply of software which is difficult to assess. In addition, it is not easy to compare the programs and to evaluate their suitability because it is first necessary to develop criteria for evaluation and to gather the necessary information. A clarification of the market is, however, not only of interest to the prospective buyer, but also for qualified dealers and for software producers wishing to enter the market and who want to assess their chances and to see where gaps exist.

In this chapter the results of an empirical examination of the authors (DYCKHOFF et al. 1988) are summarized. 28 program packages are compared which are on offer in West-Germany as standard software for the support and optimization of industrial cutting processes (cutting software). The comparison uses various evaluation criteria which are of importance to a potential user.

II. The Software Under Examination

Standard software is usually taken to mean those computer programs which can be transferred within a group of comparable users (user problems) without any fundamental alterations. All necessary modifications are carried out by setting

Table 13.1. Companies offering standard software for cutting stock problems

Company	Program
Albat + Wirsam Software-Vertriebs-GmbH	OPTAL PANOPT STAOPT II TRAOPT XOPT
CAIB GmbH	VEROS
Exapt NC Systemtechnik GmbH	NESTEX
HOLZMA-Maschinenbau GmbH	SCHNITT - PROFI(T)
Kanal u. Weissinger GmbH	OPUS-SPLAN-
KED Kerntechnik, Entwicklung, Dynamik	MINICUT
Krupp Atlas Datensysteme	Krupp ASCO - BR Krupp ASCO - CU Krupp ASCO - FR Krupp ASCO - NI Krupp ASCO - RE Krupp ASCO - ST
MAS GmbH	MASter TRIM
K + S Gesellschaft für angewandte Programmierung GmbH	KS-OPT
Dr. Ing. **Lühr** Unternehmensberatung	OPTIPROF OPTISCHNITT / GLAS OPTISCHNITT / S
Ingenieurbüro **Neugebauer**	TOP
Paul Maschinenfabrik GmbH&Co.	MAXI X.3
ORDAT GmbH & Co. KG	FALT
Schelling & Co. Maschinenfabrik und Eisengießerei	CUT PLANNER HV - HP
Siemens AG, Bereich Datentechnik	WPO CUT

parameters. A sharply defined demarcation between standard software and software for a particular user (individual software) can, however, only be achieved with difficulty, simply because of the ambiguity of the term "comparable users (user problems)". Basically, the standard set by a software is determined by the extent of the variations for its use on the basis of the various possibilities as to the choice of parameters. As a pragmatic solution, here all programs were accepted as standard which were announced as such by their producers.

At the same time, the following analysis and survey is limited to software for which we had gathered enough utilizable data in the empirical examination (Dyckhoff et al. 1988) forming the basis of this chapter. This means a total of 28 programs from 15 different companies as shown in Table 13.1.

III. A Survey of Cutting Software

The basis for the comparison of cutting software is built firstly by criteria gained from the conclusions reached in an earlier comparison of real world cutting stock problems (cf. Chapter 12 of this book). The classification criteria developed are extended and modified here with regard to a specific approach for software examination. Secondly, the usual characteristics as displayed in other software comparisons were referred to.

The products were contrasted within three survey matrices (Tables 13.2 – 13.4) using the following symbols:

● = criterion fulfilled,
○ = criterion not fulfilled,
– = criterion not applicable.

Where no utilizable data could be gathered in the empirical examination the corresponding position in the matrix is left open. The following abbreviations are used in the rows "additional services" and "documentation":

inst = installation	mod = modification
train = (introductory) training	systfl = system flow chart
maint = maintenance	progfl = program flow chart
structo = structogram	I/O = I/O formula
fdesc = file description	error = error list
opdir = operator directives	guide = user's guide.
test = test data	

The corresponding entries mean that the producers can make each service available. Where these are part of the standard program equipment (at no extra charge) this is made clear by the use of heavy type.

The most important criterion for differentiating in cutting stock problems is without doubt the dimensionality. Cutting stock problems are classified as one- or two-dimensional with regard to those dimensions relevant to the (cutting) problem. Because the dimensionality is decisive for the modelling, and thus affects the solution process, it is the clearest separating criterion for software as well, which can therefore be divided into software for one-dimensional problems and software conceived for two-dimensional problems.

1. One-Dimensional Cutting Software

It is obvious that the one-dimensional cutting programs (Table 13.2) are oriented towards specific branches of industry. Solutions are offered in the main for the materials wood, paper, and (corrugated) cardboard.

An assessment of the criterion "input figures" (shapes) is affected by the fact that very few of the programs contain information either on the maximum number of input figures (i.e. raw materials of the same form, size, orientation and quality) which could be managed at all, or which in particular, could be taken into consideration in an optimization run. This is all the more regrettable as the usefulness of software is always dependent, among other things, on the

Table 13.2. One-dimensional cutting software

		STAOPT II	MINICUT	MASterTRIM[1]
Material	used / developed for ...	wood (esp. working sheets)	tubes	paper rolls foil, sheet metal
Input figures	max. number. at all number in optimization availability limits	100 ● 	 O	 ●
Order figures	max. number. at all number in optimization tolerances	100		 ●
Objectives	trim-loss minimization cost minimization cutting costs	● O O	● O O	● ● ●
Organiza-tional adaptability	master file data order entry by system interaction	● ● ●	● ● ●	● ● ●
	data output	lists		statistics, lists, labels
Planning flexibility	isolated cutting planning	●	●	●
	additional functions	inventory, NC	order	order inventory, disposition, others
	interfaces			BDA, mainframe
Software engineering	modularity operating system program language(s)	● PC-DOS Fortran, Cobol	O MS-DOS Fortran	● PC-DOS (Turbo)-Pascal
Hardware require-ments	mainframe/medium size microcomputer memory capacity IBM PC compatibility	● ● 128 KB ●	O ● 256 KB ●	● ● 512 KB ●
Reliability	first installation installations (world/FRG) reference installation	1982 2 / 2 O	● ● ●	1984 7 / 6 ●
User friendliness	dialog oriented in German language	● ●	● ●	● ●
	additional services	inst, mod, train, maint		inst, mod, train, maint
	documentation	systfl, progfl, fdesc, I/O, opdir, error, guide		systfl, structo, progfl, guide
Costs	time unlimited usufruct	from **42 000**	**15 000**	from **90 000**

1) Five specifications: COIL TRIM, CROSS TRIM, TAMBOUR TRIM, SHEET TRIM, SHEET DOUBLE TRIM.
2) Production program outlines, paper requirements, adjustment lists.
3) Capacity planning, material requirements.

Table 13.2 (continued)

KS - OPT	OPTIPROF	TOPP	MAXI X.3	WPO
corrugated cardboard	bars, profiles	corrugated cardboard	planks bars	corrugated cardboard
●	20 20 O	arbitrary 6 O		arbitrary
●	1000 1000 O	arbitrary 50 ●	250 250 ●	arbitrary all ●
● ● O	● O O	● ● ●	● O O	● O O
● ● ●	● ● ●	● ● ●	● ●	● ● ●
lists, graphics	lists, labels	lists, other reports2)	lists	lists
●		O	O	O
	order, production	prod. planning, disposition, others3)	NC, statistics	
order, board machines, mainframe	program FENSTER, NC	order, machine control, others3)	order	program MPSI
O PC-DOS Fortran, Turbo-Pascal	● MS-DOS Basic	● PC-DOS Cobol	O Assembler	BS 2000 Fortran, Assembler
●	●	●	O	●
● 512 KB ●	● 256 KB ●	●4) - -	● O5)	O
1978 35 / 20 ●	1982 5 / 5 O	1975 75 / 9 ●	1980 100 / 30	
● ●	● ●	● ●	● ●	● ●
inst, mod, **train**, maint	**inst**, mod, **train**, maint	**inst**, mod, **train**, maint	**inst**, mod, train, maint	
systfl, progfl, guide	**guide**	**opdir**, guide (on screen)	systfl, structo, progfl, fdesc, I/O, **opdir, error**, test, **guide**	
77 000	**5 000**	**55 000**	6)	**15 000***

4) In preparation (for PC-DOS and OS/2 and language C; IBM-compatibility.
5) Own development: PAUL CNC-control.
6) Sold only with mechanical equipment.

size of the problem, and the number of input figures to be considered does have a considerable influence on the problem size. The quality of the solution is further dependent on whether the program falls back on a limited number of the figures in optimization (chosen, for example, according to given priority rules). In addition to this, one-dimensional cutting software does not usually take into account availability limits for input figures (for example, remainders in stock).

Analogous to the input figures, the criteria "maximum number at all" and "number in optimization" as regards the order figures cannot be uniformly judged because of the lack of sufficient information. Tolerances in carrying out orders, in the sense that a certain amount of overproduction or underproduction of the orders can be permitted in order to improve the optimization results, can be found in several programs.

Objective of the cutting planning mostly is the mere minimizing of waste (trim loss minimization). Only a few products aim as well at minimizing costs, in particular taking those cutting costs into consideration which are caused by the production process itself.

The organizational adaptability, i.e. the ability to fit into the existing organizational structure of a firm, seems to be satisfactory in all the products. It is usual to have master file data collection via parameters and recording commission data by the system (order entry by system); the choice of cutting patterns can either take place automatically or can be influenced by a possible interaction through the planner.

In contrast, the planning flexibility of many one-dimensional programs seems to be insufficient. The performance of software depends to a great extent on its ability to integrate with planning and other areas within a firm. The decision in favour of a software product can depend in two different ways on the manner it is fitted out as regards planning flexibility. On the one hand, it is possible that a user wants to employ the software simply for cutting planning on its own (isolated cutting planning), either because he does not need any data processing in other areas, or because he already has a separate data processing solution for other planning or functional areas. On the other hand, it is possible that the user wants to integrate cutting planning with other planning and functional areas such as order processing, accounting, inventory management, or numerical control. In this case it is essential that the software has the desired additional functions or can be extended using additional modules to include them, or is provided with the corresponding interfaces. Although it is true that almost all products have at least one connection with other planning and functional areas, mostly with inventory management, order processing or numerical control, the possibilities for connecting with other areas appear to be insufficient, especially in the light of the discussion on comprehensive CIM concepts. Interfaces with accounting and other business departments would be desirable.

The software engineering quality of each system varies extremely. Not all programs are built up on a modular basis and thus do not fulfil the basic precondition for a good system concept.

For most potential users in small and medium-sized firms it is probably of advantage that nearly all the programs are offered in an IBM PC compatible version (hardware requirements).

Software products must be reliable in that they must deliver correct results for all input combinations and must be breakdown-proof. It is extremely difficult to obtain any guaranteed user information concerning the exactness of a program and whether it is proof against breakdown. At most the number of installations (worldwide/FRG) in relation to the date of the first installation, and the information as to reference installations can be indicators in this respect. The details regarding these points seem to show that not all programs are (as yet) fully mature.

User-friendliness is regarded by users as one of the most important software characteristics. This criterion is, however, difficult to pin down as it depends to a great extent on the individual requirements and competence of each user. Indicators regarding user-friendliness are: dialog oriented programs in German language, extensive (German-language) program documentation, the readiness of the software producers to support the users by means of additional services such as program maintenance and introductory courses. As regards these indicators the only significant difference can be seen in the documentation on offer. Otherwise the programs are, without exception dialog-oriented with German language user guides. All producers support the users in program implementation, carry out modifications, and offer training and maintenance, though some charge extra for these services.

2. Two-Dimensional Cutting Software

In the case of two-dimensional cutting software the form of the cut products plays a major part because of the differences in the cutting technologies involved. Basically, a distinction has to be made between rectangular cutting only and cutting arbitrarily shaped forms.

(a) Rectangle Programs

The construction of (cutting) patterns depends on the existing cutting technology, where guillotine and nested cuts must be distinguished. The examined rectangle programs (Table 13.3) mostly support cutting processes with guillotine cuts where – provided that information exists – pattern construction can take up to three-stage cuts into consideration (number of stages). At the same time nearly all the programs make allowances for taking the structure of the material into account (orientation) where this is necessary, e.g. for veneers and panels for decoration.

The rectangle programs hardly show any differences to the one-dimensional programs as regards the criteria "order figures", "objectives", "organizational adaptability", and "user-friendliness".

Differences do, however, exist in that nearly all rectangle programs enable availability limits to be entered, the number of installations tends to be higher, and information regarding reference installations is found more often. The programs also appear to be better equipped as regards planning flexibility. All the products examined are connected to order processing and numerical control,

Table 13.3. Rectangle programs

		OPTAL	PANOPT	TRAOPT	XOPT	VEROS
Material	used/ developed for ...	sheet metal, wood, synthetics	chipboard, fibreboard	float glass	float glass	carpet and other rolls, steel coils
Pattern	nested cuts	O	●		O	O
	number of stages					3
	orientation	●	●	O	●	●
Input figures	max. number at all		10		10	arbitrary
	number in optimization	10	10	10		all
	tolerances	●	●	●	●	●
Order figures	max. number at all		88	256	256	arbitrary
	number in optimization	99				all
	tolerances	O	●	O	●	●
Objectives	trim-loss minimization	●	●	●	●	●
	cost minimization	O	●	O	O	●
	cutting costs	O	●	O	O	O
Organiza-tional adaptability	master file data	●	●	●	●	●
	order entry by system	●	●	●	●	●
	interaction	●	●	O	●	●
	data output	lists, graphics	graphics	lists, graphics	lists, graphics	lists, graphics
Planning flexibility	isolated cutting planning	O	●	●	●	●
	additional functions	inventory, order	NC, management of standard plans			
	interfaces	NC	inventory, order, NC	order, NC	NC, order, contour program	order, NC
Software engineering	modularity	O	●	O	●	●
	operating system	PC-DOS	PC-DOS	PC-DOS	PC-DOS	138 CPF
	program language(s)	Fortran, Basic, Assembler	Fortran, Cobol, Assembler	Basic, Cobol, Assembler	Fortran, Basic, Assembler	RPG III
Hardware require-ments	mainframe/medium size	●	O	●	●	●
	microcomputer	●	●	●	●	O
	memory capacity	320 KB	512 KB	128 KB	256 KB	-
	IBM PC compatibility	●	●	●	●	-
Reliability	first installation	1982	1984	1976	1983	1986
	installations (world/FRG)	8 / 6	40 / 20	300 /	98 / 29	1 / 1
	reference installation		●			●
User friendliness	dialog oriented	●	●	●	●	O
	in German language	●	●	●	●	●
	additional services	inst, mod, train, maint	inst, mod, train, maint	inst, mod, train, maint	inst, mod, train, maint	inst, mod, train, maint
	documentation	systfl, progfl, fdesc, I/O, opdir, error, test, guide	opdir, guide	systfl, fdesc, I/O, opdir, error, test, guide	fdesc, opdir, error, test, guide	systfl, fdesc, I/O, opdir, (guide)
Costs	time unlimited usufruct	42 000	from 45 000	from 25 000	from16 500	from 29 500

1) MS DOS, VMS, Primos, UNIX, AEGIS.
2) Order information, material reports, sawing times and costs, remainder lists.

Table 13.3 (continued)

SCHNITT-PROFI(T)	Krupp ASCO-RE	OPTI-SCHNITT/GLAS	OPTI-SCHNITT/S	FALT 4)	CUT PLANNER	HV - HP	CUT
all flat material	wood, sheet metal, paper, glass, synthetics	float glass	flat material	cardboard, boxes	chipboard, fibreboard, synthetics	chipboard, fibreboard, synthetics	boards,rolls, profiles
● 3 ●	● 4 ●	○ 3 ○	○ 3 ●	● - ●	○ 3 ●	○ 3 ●	● unlimited ●
25 25 ●	400 ●	10 10 ●	1000 x 10 10 ●	○	10 10 ●	10 10 ●	20 20 ●
120 ●	400 ●	99 x 250 250 ○	99 x 250 250 ○	○	50 50 ●	60 60 ●	100 100 ●
● ● ●	● ● ●	● ○ ○	● ○ ○	● ○ ●	● ○ ○	○ ● ●	● ○ ●
● ● ●	● ● ●	● ● ●	● ● ●	● ● ●	● ● ○	● ● ●	● ○ ●
lists, graphics	lists, graphics	lists, graphics, labels, others	lists, graphics, remainder lists, others		graphics,labels, others 2)	graphics,labels, others 3)	lists, graphics, statistics
●	●	○	○	○	●	●	●
	remainders, BDA, NC, standard plans	order, production	order, inventory, remainder management	cost estimate	NC, order	NC, order	
Host	inventory, order, NC	NC, order	joinery program, NC	order			
○ MS-DOS Basic, Cobol	● various 1) Fortran	● MS-DOS	● MS-DOS	○ Basic	● MS-DOS Cobol	● RTE-A / MS-DOS Fortran	● Fortran
○	●	●	●	●	○	●	●
● 512 KB ●	● 512 KB ●	● 256 KB ●	● 256 KB ●	○ - -	● 256 KB ●	● 512 KB ●	○ - -
1983 150 / 12 ●	1976 59 /21 ●	1976 36 /26	1982 9 / 8	1979 10 / 10 ●	1982 500 / 20 ●	1977 15 / 6 ●	
● ●	● ●	● ●	● ●	● ●	○ ●	● ●	● ●
inst, mod, train, maint	inst, mod, train, maint	inst, mod, train, maint	inst, mod, train, maint	inst, mod, train, maint	mod, train, maint	inst, mod, train, maint	
fdesc, opdir, test, guide	progfl, fdesc, I/O, opdir, error, test, guide	fdesc, test,guide	systfl, fdesc, I/O, opdir, error,test, guide	systfl, fdesc, I/O, error, guide	fdesc, I/O, opdir, error, guide	fdesc, I/O, opdir, error, guide	
22 043	from 30 000	from 6 000	8 000	22 000	2 000 - 20 000		15 000

3) See 2); in addition: sorting and stacking plans, shift reports, costumer reports.
4) Although the program FALT displays a few characteristics of the "contour programs", on the whole it belongs to the group of the "rectangle programs".

Table 13.4. Contour programs

		NESTEX	OPUS - SPLAN
Brief description		system for all nesting tasks in oxy-acetylene cutting, wire drawing, plasma-arc cutting, laser cutting, nibbling, stamping	nesting program for oxy-acetylene cutting
Material		all flat material, esp. sheet metal	sheet metal
Functions		automatic and interactive nesting plans for trim-loss minimal cutting of (metal) sheets (utilization of inner surfaces, guaranteed overlap-free)	contour-delineation via geometric processor, automatic nesting plans, plotter outputs
Planning flexibility		interfaces to PPS- and CAD-systems, NC	output in format requested by control system
Software engineering	modularity	●	●
	operating system	VMS, Primos, DOS	MS DOS
	program language(s)	Fortran 77	Pascal
Hardware requirements	mainframe/medium size	●	O
	microcomputer	●	●
	memory capacity	2 MB	512 KB
	IBM PC compatibility	O	●
Reliability	first installation	1982	1983
	installations (world/FRG)	13 / 10	
	reference installation		●
User friendliness	dialog oriented	●	●
	in German language	●	●
	additional services	**inst**, mod, **train**, maint	inst, mod, **train**, maint
	documentation	**systfl, fdesc, error, test, guide**	**structo, fdesc, error, guide**
Costs	time unlimited usufruct	from **35 000**	

Table 13.4 (continued)

Krupp ASCO				
describes a standard program family; automatic nesting plan construction for all current and NC controlled separating processes; extra version for rectangular geometry (cf. Tab. 13.3)				
ASCO - BR nesting plans for oxy-acetylene cutting	**ASCO - CU** nesting plans for cutting compound products	**ASCO - FR** nesting plans for CNC-boring and milling centres	**ASCO - NI** cutting plans for nibbling/stamping/laser autom. machines	**ASCO - ST** cutting graphics for stamping from (single/double) coil
sheet metal	fibre-stringthen-ed compounds	sheet metal	sheet metal	sheet metal
- order and inventory management - assumption of geometric description (e.g. from NC programming system) - automatic nesting taking cutting technology into account - manual subsequent nesting - fixing of cut sequence - output of cutting plan as plot or list - output of NC program for the whole cutting plan				
modular construction enables combinations of individual systems up to integrated solutions for fully-automatic production process				
● OS/MVS, VMS,Primos Fortran	● MVS, VMS Fortran	● AEGIS, VMS, Org-M Fortran	● VMS, UNIX Fortran	● OS/MVS, VMS Fortran
● O - -	● O - -	● O - -	● O - -	● ● - -
1976 33 / 15 ●	1985 2 / 2 ●	1982 4 / 2 ●	1984 4 / 4 ●	1982 4 / 4 ●
● ● **inst**, mod, **train**, maint	● ● **inst, mod, train, maint**	● ● **inst**, mod, **train**, maint	● ● **inst**, mod, **train**, maint	● ● **inst**, mod, **train**, maint
progfl, fdesc, opdir, error, test, guide	**fdesc, I/O, opdir, error, test, guide**	**fdesc, I/O, opdir, error, test, guide**	**systfl, structo, progfl, fdesc, opdir, error, test, guide**	**fdesc, I/O, opdir, error, test, guide**
from **70 000**	from **120 000**	from **120 000**	from **70 000**	from **55 000**

most of them are also connected to inventory management. A further obvious difference from the one-dimensional cutting software is that it contains considerably more products with a modular construction. Finally it must be said that many programs are also on offer in a mainframe version, but nearly all the programs can be run on microcomputers.

(b) Contour Programs

The two-dimensional cutting software for arbitrary contours (Table 13.4) has a special place in the range of cutting software on offer. These programs are, without exception, nesting programs which take the technical production requirements of the cutting process into account. In order to do justice to these special requirements, Table 13.4 emphasizes points different from the other tables.

All the programs have a strictly modular structure. Therefore, the possibilities for their use range from isolated cutting planning to computer-integrated manufacturing systems.

What is particularly striking about this software group is the fact that many of the programs were conceived exclusively for sheet-metal working. This is probably because production in this branch is often highly automated.

IV. Summary

There is far less software on offer for one-dimensional cutting stock problems and for two-dimensional problems with arbitrary contours than there is for two-dimensional problems with rectangular geometry, and here programs for guillotine cuts only dominate the market. Furthermore, the first two groups concentrate on just a few types of material whereas the rectangle programs on offer cover a whole range of branches.

Trim loss minimization is part of the standard offer but cost minimization is found far less often, although cutting costs are then usually taken into account as well.

As far as their organizational adaptability is concerned nearly all the programs achieve good results. With reservations, the same applies to user-friendliness. There are, however, great differences where planning flexibility is concerned. Existing interfaces and additional functions in the first place refer to order processing and numerical control, and less to inventory management. All the same there are some programs which could be part of a CIM system.

Most of the programs can be used on IBM PCs with the usual capacity and operating systems. For this reason they are of particular interest to small and medium-sized firms. Software for mainframe computers or medium-sized data technology is found more seldom. The programs are usually written in one of the common higher-level programming languages, whereby there is a clear difference regarding modularity between the one-dimensional and the two-dimensional cases.

None of the software is much older than ten years; in most cases the first installation is of a later date. Sometimes this will explain why there has been up to now a relatively low number of installations. On the other hand, some of the newer programs have had a large number of installations worldwide; this no doubt shows that they were imported from abroad and have been in use for some time there. A drawback of many programs, especially of the one-dimensional sort, is their lack of reference installations.

If the number of reference installations is taken as a yardstick, it can be said that up to now only a few of the programs have been really successful. These are concentrated among still fewer sellers (cf. Table 13.1) and are to be found in the higher price classes. Some of the sellers are, however, new to the market, and only time will tell whether they are successful or not.

The prices range from 6,000 to 120,000 DM for the right to use a program with no time limit, whereby the installation costs, the costs of the introductory support and training, when offered, are often included in the price; modification costs are hardly ever included, and maintenance costs almost never.

V. Concluding Remarks

It was not our intention to carry out a comparative rating of the cutting software we examined, mainly for two reasons. Firstly, the reliability of the data in Tables 13.2 – 13.4 can be questioned, as up to now the data have been based solely on information supplied by the sellers and call for careful examination. All the same, an interested user can make a rough choice using the comparison of the programs presented here. Of particular importance are probably the "knock out criteria": costs, dimensionality, hardware requirements. Furthermore, the other criteria enable a detailed comparison to be made of programs chosen beforehand and made available by the sellers by using test runs made under realistic conditions.

The second reason follows directly from this. Earlier examinations and further experience show that cutting stock problems in practice differ one from another a great deal, even when they are apparently similar. A comparative rating of cutting software can therefore nearly always be made as to a concrete situation, only, and can vary according to the conditions for use. This type of scrutiny has not yet been carried out. All the same, the results of our examination give the impression that the two-dimensional cutting software products on offer – and especially the contour programs – are much more developed than the one-dimensional products.

Reference

Dyckhoff, H., Finke, U., and Kruse, H.-J.: Empirische Erhebung über Verschnittsoftware, in: Operations Research-Spektrum, 1988, issue 4.

List of Contributors

DIETER ABEL
Getriebebau-Nord
D-2072 Bargteheide
Federal Republic of Germany

HARALD DYCKHOFF
Dept. of Production Management
RWTH Aachen
Templergraben 64
D-5100 Aachen
Federal Republic of Germany

GÜNTER FANDEL
Dept. of Production Management
FernUniversität
Feithstr. 140
D-5800 Hagen
Federal Republic of Germany

UTE FINKE
Dept. of Operations Research
FernUniversität
Feithstr. 140
D-5800 Hagen
Federal Republic of Germany

PETER FRANÇOIS
Dept. of Production Management
FernUniversität
Feithstr. 140
D-5800 Hagen
Federal Republic of Germany

TOMAS GAL
Dept. of Operations Research
FernUniversität
Feithstr. 140
D-5800 Hagen
Federal Republic of Germany

HERMANN GEHRING
Dept. of Economics
Freie Universität Berlin
Corrensplatz 2
D-1000 Berlin 33
Federal Republic of Germany

MANFRED GRAUER
Dept. of Computer Science
Universität Dortmund
POB 500500
D-4600 Dortmund 50
Federal Republic of Germany

HERMANN-JOSEF KRUSE
Dept. Applied Mathematics
Schering AG
Müllerstr. 170 – 178
D-1000 Berlin 65
Federal Republic of Germany

EUGEN MAY
Dept. of Production Management
FernUniversität
Feithstr. 140
D-5800 Hagen
Federal Republic of Germany

JOACHIM REESE
Dept. of Management
Universität Bonn
Adenauerallee 24 – 42
D-5300 Bonn
Federal Republic of Germany

JOHANNES WOLF
Am Freibad 5
D-5800 Hagen
Federal Republic of Germany

MIX
Papier aus verantwortungsvollen Quellen
Paper from responsible sources
FSC® C105338

If you have any concerns about our products,
you can contact us on
ProductSafety@springernature.com

In case Publisher is established outside the EU,
the EU authorized representative is:
Springer Nature Customer Service Center GmbH
Europaplatz 3, 69115 Heidelberg, Germany

Printed by Libri Plureos GmbH
in Hamburg, Germany